NANOTECHNOLOGY SCIENCE AND TECHNOLOGY

NANOCOMPOSITE COATINGS

NANOTECHNOLOGY SCIENCE AND TECHNOLOGY

Additional books in this series can be found on Nova's website
under the Series tab.

Additional E-books in this series can be found on Nova's website
under the E-book tab.

MATERIALS SCIENCE AND TECHNOLOGIES

Additional books in this series can be found on Nova's website
under the Series tab.

Additional E-books in this series can be found on Nova's website
under the E-book tab.

NANOTECHNOLOGY SCIENCE AND TECHNOLOGY

NANOCOMPOSITE COATINGS

MAHMOOD ALIOFKHAZRAEI
AND
ALIREZA SABOUR ROUHAGHDAM

Nova Science Publishers, Inc.
New York

For permission to use material from this book please contact us:
Telephone 631-231-7269; Fax 631-231-8175
Web Site: http://www.novapublishers.com

Additional color graphics may be available in the e-book version of this book.

LIBRARY OF CONGRESS CATALOGING-IN-PUBLICATION DATA

Aliofkhazraei, Mahmood.
Nanocomposite coatings / authors, Mahmood Aliofkhazraei and Alireza Sabour Rouhaghdam.
p. cm.
Includes index.
ISBN 978-1-61122-138-1 (hardcover)
1. Protective coatings. 2. Nanocomposites (Materials) I. Rouaghdam, Alireza Sabour. II. Title.
TA418.76.A45 2010
667'.9--dc22

2010037362

Published by Nova Science Publishers, Inc. † New York

CONTENTS

PREFACE

This book reviews research activities around fabrication of these kinds of two dimensional nanostructured coatings with examples of enhanced mechanical properties, corrosion resistance, and physical characteristics. As one of the useful and simple methods for fabrication of nanocomposite coatings, electrochemical deposition (electroplating) techniques are a strong focus in this book. The relation among nanotechnology and these kinds of nanostructures that come from "size effect" and "distribution effect" is discussed through different chapters of this book. Nanocomposite coatings have numerous advantages.

Chapter 1

Synthesis and Processing of Nanocomposite Coatings

Abstract

This chapter discusses an introduction about nanocomposite coatings and their fabrication methods. It starts from surface engineering and goes through different technologies for synthesis and processing of nanocomposite coatings. Detailed discussion about electrodeposition method was presented in this chapter and fabrication of metallic matrix nanocomposite coatings was reviewed. Initial discussions about the affected characteristics of nanocomposite coatings were written with respect to un-composed coatings. Considered requirements for surface were: increasing the strength against friction, abrasion, and corrosion, or boosting the thermal, optical, magnetic, and electrical properties. Using materials for improving surface properties is not economically advised. Thus, for high rate of yield it is recommended to use a sub-layer with efficient properties, and materials which are cheaper and easier to reshape. Ideally, sub-layer must be optimized for maximizing the coating benefits and, consequently, creating the most efficient coating system. Hence, the final product eliminates the need for rare materials and can be a cheaper and of a better yield solution, compared with early solutions.

1.1. Surface Engineering

Surface engineering can be defined as an enabling technology used in a wide range of industrial activities. Till two past decades this technology was not practically known. The subject of this phenomenon as independent branch - involving engineering science, physics, and material science - was affected to a high extent by recent progresses in the field of coating and recognizing that surface is the most important part of each engineering block.

Surface engineering was founded by detecting surface features which destroy most of pieces, e.g. abrasion, corrosion, fatigue, and disruption; then it was recognized, more than ever, that most technological advancements are constrained with surface requirements. Considered requirements for surface were: increasing the strength against friction, abrasion, and corrosion, or boosting the thermal, optical, magnetic, and electrical properties. For instance, fuel yield and specific output power of thermal engines, such as gas turbines or adiabatic diesels, is limited by hot corrosion and the properties of thermal barriers of special pieces of surfaces. Similarly, in a wide range of industry (such as nuclear power, gas and oil exploitation, mining, and manufacturing), the surface generates an important problem in technological advancement. Using total material for improving surface properties is not economically advised. Thus, for high rate of yield it is recommended to use a sub-layer with efficient properties, and materials which are cheaper and easier to reshape. Ideally, sub-layer must be optimized for maximizing the coat benefits and, consequently, creating the most efficient coating system. Hence, the final product eliminates the need for rare materials and can be a cheaper and of a better yield solution, compared with early solutions.

This just elucidates an aspect of surface engineering: products improving. Probably, the key importance of surface engineering is producing new materials, which are just resulted in advance coating. Though electronic and optical electronic apparatuses are sharp examples, but nowadays most of mechanic engineering blocks are included in this topic. The title of surface engineering includes all methods and processes which are applied to cope, modify, and raise the surface yield, e.g. strength against abrasion, fatigue, abrasion, corrosion, and adjusting with environment. It has three topics, including:

1) Optimizing surface properties and surface (and sub-layers) yield in terms of corrosion, adhesion, abrasion, and the other mechanical and physical properties.
2) Coating technology includes traditional technologies such as coloring, plating, weld surfacing, plasma spray, high speed spray, variant thermal and thermo-chemical operations - such as nitriding and Carburizing - in addition to recent technology in surface operation with laser, physical and chemical deposition of vapor, and ion implementation.
3) Describing coating characteristics including surfaces and interfaces evaluation including chemical composition, morphology, and mechanical, electrical, and optical structures and properties.

1.1.1. Coating Technology

General categorization of surface engineering techniques and their brief comparison are show, respectively, at tables 1.1 and 1.2.

1.2. Electrodeposition

Electroplating and electrodeposition are performed through electrical deposition process, which is indeed a metal coating deposit created by electrical current application. Electroplating is carried out for improving surface traits of mainly metallic blocks, for achieving the following favorable features:

a) Protection against corrosion
b) Resistance against abrasion and friction
c) Creating electrical resistivity or conductivity
d) Creating thermal barrier
e) Decorative cases

At electrical deposition, according to figure 1.1, negative charge is applied on covered block surface, using an external voltage. Then coatings are developed by submerging the pieces into a solution - rich by metallic ions - and applying the electrical current through that. Metallic ions are provided by either metallic salts dissolution into plating solution (electrolyte) or positive pole (anodes) electrode dissolution during plating. In this condition, due to potential difference, positive ions (cations) move toward cathode and reduce on its surface as metallic atoms.

At electrical deposition there must be an electrical circuit, in the presence of a battery or supply resource – far from charged ions and electrolyte in solution. Thus, for performing electrical deposition process the circuits' electrical conductivity and metallic ions motivation is necessary.

Applying the electrical conductivity of circuit, sub-layer's surface must include no surface pollutions or oxides. In this case metallic ions mobility increases through heating and lack of impurities in electrolyte. Hence, effective factors on coatings developed by electrical deposition are as following:

a) Applied current density
b) Plating solution chemical composition
c) Temperature of plating solution
d) Physical and chemical properties of sub-layers' surface

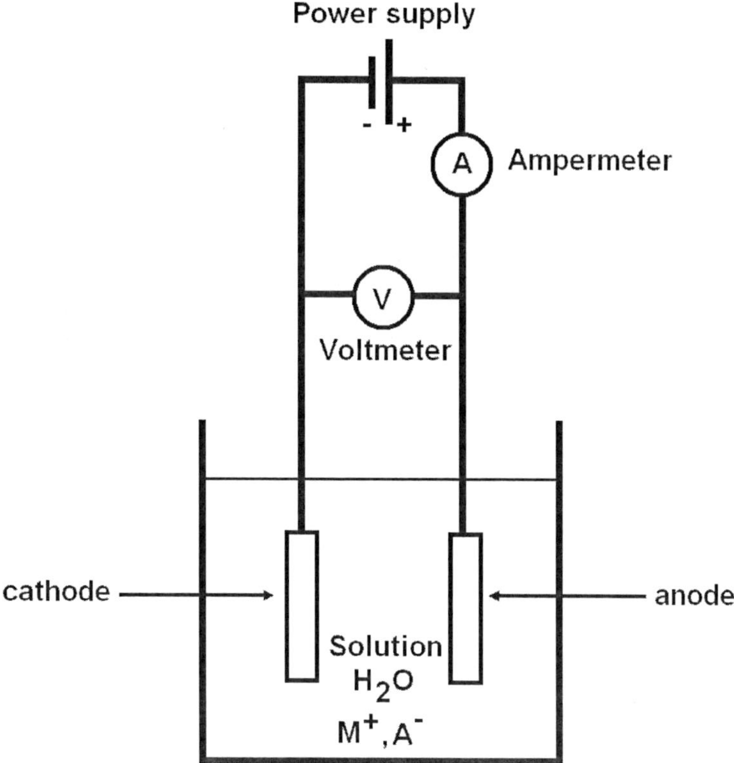

Figure 1.1. Schematic set up for electrodeposition.

Table 1.1. General classification of surface engineering methods

Surface engineering										
Surface treatments				Surface coatings						
Finishing	Thermal	Thermo-chemical (cvd)	Ion implantation	Chemical	Electro deposition	Physical vapour deposition	Chemical vapour deposition	Thermal spraying	Welding	Cladding
Grinding	Flame hardening	Boriding	Nitrogen implantation	Sol gel	Electroless plating	Sputtering	Atmosphering	Air plasma	Gas	Friction surfacing
Machining	Laser treatment	Nitriding		Anodising	Electro-pulse plating	-diode	Loe pressure	Vaccum plasma	-rod	Electro spray deposition
Blasting	Electron beam treatment	Carburising		Electro-phoresis	Occluded coating	-magnetron	Laser	D-gun	-powder	Brazing
Peening	Induction hardening	Aluminising		Electrostatic spraying		-ion beam	Hot filament	High velocity flame	-paste	Diffusion bonding
Rolling		Chromising		Paint/slurry spraying		-triode	Metallorganic	Water stabused	Arc	Explosive cladding
		Plasma variants				-Rf/dc	Plasma variants	Powder flame	-TIG	
						Evaporation	-Direct current	Arc spraying	-MIG	
						-Resistive	-Pulse plasma	Wire spraying	-electro slag	
						-Inductive	-Rf/ac	Inductive plasma	- submerged arc	
						-Electron beam	-Microwave	Laser	-shielded metal arc	
						-Arc			Laser	
						Plasma variants				
						-Ion plasma				
						-Pulsed plasma				
						Ion assisted coating				

Table 1.2. Comparison among different surface engineering methods in gas, liquid and semi-liquid phases

Gaseous (vapour) phase			Liquid phase			Molten, semi-liquid phase		
	PVD	CVD	Ion implantation	Sol-gel	Electro-deposition	Laser	Thermal spraying	Welding
Thickness (mm)	Up to 3	Up to 0.5	-0.005	0.002	0.02-0.5	0.05-2	0.05-2	1-2 or more
Deposition rate ($kg.h^{-1}$)	Up to 0.5 per source	Up to 1	-	0.1-0.5	0.1-0.5	0.1-1	0.1-10	3-50
Component size	Limited by chamber size			Limited by solution bath		No practical limits		
Substrate material	Wide choice	Limited by deposition temperature	Weak choice	Weak choice	Weak choice	Weak choice	Weak choice	Mostly steels
Pre-treatment	Chemical plus ion bombardment	Various	Chemical plus ion bombardment	Grit blast and/or chemical clean	Chemical cleaning and etching	Mechanical and chemical cleaning		
Post-treatment	None	Substrate stress relief/mechanical properties	None	High temperature calcine	None/Substrate stress relief			
Control of deposit thickness	Good	Fair/Good	Good	Fair/Good	Fair/Good	Manual- Variable Automated- Good		
Bonding mechanism	Atomic	Atomic	-	Surface forces		Mechanical/Chemical		Metallurgical
Distortion of substrate	Low	Can be high	Low	Low	Low	Low/Moderate		Can be high

1.2.1. Advantages and Disadvantages of Electroplating

Electroplating advantages are:

1) As operation temperature does not exceed over 100 °C, work-piece is not subject to any metallurgic deformation or unwanted changes.
2) Adjusting the plating condition one can change the solidity, internal stress and metallurgic characteristic of the coating.
3) The obtained coating through this method has an efficient compaction and adhesion with its sub-layer. In this method the bound nature is molecular and might reach up to 1000 MPa.
4) The coating thickness is adjustable through a change in applied current density and the length of plating.
5) In this case, there are no technical limitations in coating thickness; and for metals such as nickel, we can obtain the thickness of 13mm by electroforming and reclamation.
6) Creating this kind of coating, in some extent, is not affected by line of sight factor. Although the throwing power (i.e. the ability of coating at rounded corners) may be limited, but there is a relative freedom in anodes arranging (e.g. in slim tubes interior coating).
7) We can easily cover the surfaces which has no need for coating.
8) The process is efficient for automation.

Disadvantages of plating are:

1) As current density distribution is not equal in the piece, the coating intends to be thickened at corners and edges and be thinned at dents and center of flat broad surfaces.
2) Deposition rate rarely exceeds over 75μm/h; however through electrolyte circulation it can be accelerated.
3) The size of plating bath is limited by work-piece dimensions.

1.2.2. Electroplating History

History of applying electroplating comes back to 1800 when Brugnaltelli, an Italian chemist which also was a university professor, coated the gold for fist time using the electroplating. Since he was a friend of Volta, inventor of electrical voltaic batteries, he had a chance to work with different metallic plating solutions using the voltaic electricity. As the current European scientific community did not support him, his work remained highly unknown out of Italy. It was the same till 1839, when Russian and English scientists worked independently in deposition processes - similar to that of Bruagnatelli's – on copper plating at printer machine sheets. In 1840, Elkington and Henry in cooperation with Wright – the inventor of sciaenid plating bath composition - recorded the first electroplating of the gold and silver. Nowadays, through modern plating bathes with new compositions one can perform continuous plating of cords and belts, semi-conductors, and also the metallic pieces with complex shapes.

Electroplating can be categorized into four groups, in terms of developed phases in the deposited metals, including:

1) Single metal plating
2) Alloy plating
3) Composite plating
4) Alloy-composite plating

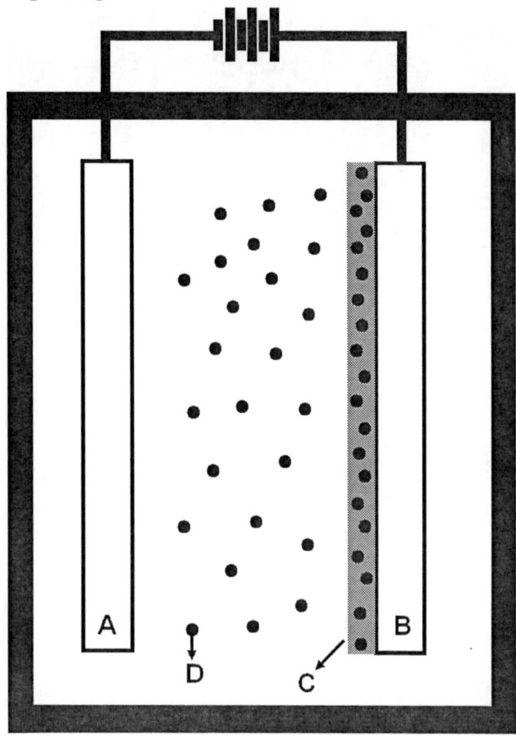

Figure 1.2. Schematic figure of co-electrodeposition (A: anode, B: cathode, C: matrix, D: micro- or nanoparticles).

1.3. Composite and Nanocomposite Plating

Composite plating, which is also addressed as electrochemical composite plating, is among the methods for creating composite materials with metallic matrix. As it can be noticed from figure 1.2, composite plating is carried out via simultaneous electrodeposition process, which is a process of combining the insoluble dispersive particles in electrolyte during an electrical deposition of a metal. Through this process we can develop thin layers of metal matrix composites bearing particles of pure metal, ceramics and organic materials.

The particles sizes are from 2 nm to 100μm, which are generally used in the matrix of copper, nickel, cobalt, and different alloys, resulted from electrodeposition. Suspended particles concentration in plating solution varies between 2 to 200 g/lit; and one can typically produce composites with volumetric ratio of 1 to 50 from them.

1.3.1. History and Application of Composite Plating

It seems that the first electrochemical coatings – which had ceramic particles in the metallic matrix for protection against high temperature oxidation and abrasion – were developed by Fabian and Withers, respectively, in 1962 and 1963. The simultaneous electrical deposition has a key role in many technological fields such as transportation, manufacturing, power supply, and aerospace..

Particular functions of coatings developed from this process include:

- The coating of surface cutting tool
- Creating materials with high thermal conductive characteristics and dispersion consolidating for active cooling structures such as water cooler systems.
- Developing the electrodes with high surfaces to electro-catalysts for converting solar energy and operations on dangerous waste waters.
- Creating self-lubricate coatings in the rage of temperature.
- Resistance against abrasion in internal combustion engines, particularly high temperature oxidation in plane engines.
- Applying at micro electro-mechanic systems (MEMS) drivers and electronic connections.
- Coating glass shaping casts and weaving machines.

Table 1.3. Comparison between mechanical properties of WC/Ni-Co composite coating fabricated by nanocomposite electrodeposition and thermal spray

Coating method	Concentration WC (Vol. %)	Hardness (VHN)	Wear rate (mm^3/Nm)	Friction coefficient	Roughness (μm)
Electrodeposited WC/Ni-Co	25-30	850	0.4×10^{-5}	0.31	~ 0.7
Thermal sprayed WC/Co	88	~ 1050	0.1×10^{-5}	0.5	~ 3

1.3.2. Advantages and Disadvantages of Composite Plating

Composite plating advantages are:

- Simple and applicable use for pieces with any geometric shape
- Lack of problems related to high pressure and temperature processes such as thermal spray process
- Capability of producing materials with gradual properties for reducing probable phenomena of delamination and spalling.
- Capability of performing the process in a continuous manner
- Decrease in the damages which immersion and spray methods are faced with

Table 1.4. Some of physical and mechanical properties of oxide ceramics (ionic)

Material	Crystalline structure	Density. kgm^{-3}	Knoop microhardness. GPa $(kg.mm^{-2})$	Young's modulus. GPa	Poisson's ratio	Tensile strength. MPa	Compressive strength. MPa	Flexural strength. MPa
Al_2O_3	Hexagonal	3980	20.4* (2100)	390	0.23	300	3400	500-600
Cr_2O_3	Hexagonal	5210	12.7 (1300)	-	-	-	-	95
TiO_2	Tetragonal	4250	10.8 (1100)	280	-	70-100	800-1000	350
SiO_2 (fused silica)	Amorphous	2200	$5.4\text{-}7.4^{+}$ (550-750)	73	0.17	30-70	700-1400	50-120
$HfO_3\text{-}Y_2O_3$	Tetragonal (practically stabilized)	10010	7.4 (750)	-	-	-	-	350-2500
$ZrO_2\text{-}Y_2O_3$ (92-8%)	Tetragonal (practically stabilized	6100	12.7 (1300)	290	0.24	45	2000	500-700
BeO	Hexagonal	3008	14.7 (1500)	380	0.34	105	2100	275
MgO	Cubic	3580	7.4 (750)	320	0.19	130	830	300
ThO_2	Cubic	10010	6.4 (650)	240	0.28	100	1500	130

Material	Fracture toughness (K_{IC}). $MPam^{1/2}$	Thermal conductivity. $Wm^{-1}K^{-1}$	Specific heat. $Jkg^{-1}K^{-1}$	Coefficient of thermal expansion $\times10^{-6o}C^{-1}$	Electrical resistively. μΩcm	Melting or decomposition temperature/ maximum operating temperature. °C
Al_2O_3	4.9	35	920	7.1	10^{20}	2050/1800
Cr_2O_3	-	7.5	730	5.4	-	2300/1000
TiO_2	-	8.3	785	8.8	-	1840/700
SiO_2 (fused silica)	-	1.2	670	0.9	$10^{15}\text{-}10^{18}$	1650/575
$HfO_3\text{-}Y_2O_3$	-	1.7	345	5 .3	-	2840/1700
$ZrO_2\text{-}Y_2O_3$ (92-8%)	8.5-9.5	1.8	630	11	10^{18}	2760/1700
BeO	-	250	1000	7.4	10^{23}	2570/1100
MgO	-	50	880	11.6	10^{13}	2850/1700
ThO_2	-	11	250	9.5	10^{16}	3220/1700

Table 1.5. Some of physical and mechanical properties of silicide ceramics

Material	Crystalline structure	Density. kgm^{-3}	Knoop microhardness. GPa ($kg.mm^{-2}$)	Young's modulus. GPa	Poisson's ratio	Tensile strength. MPa
$TiSi_2$	Orthorhombic	4150	8.8 (900)	270	-	-
$ZrSi_2$	Orthorhombic	4880	9.8 (1000)	270	-	-
$HfSi_2$	Orthorhombic	8030	8.6 (875)	-	-	-
VSi_2	Hexagonal	5100	9.8 (1000)	-	-	-
$NbSi_2$	Hexagonal	5690	7.8 (800)	-	-	-
$TaSi_2$	Hexagonal	9140	13.7 (1400)	-	-	-
Cr_3Si_2	Tetragonal	5600	12.3 (1250)	-	-	-
$MoSi_2$	Tetragonal	6240	14.2 (1450)	420	-	150
WSi_2	Tetragonal	9870	10.8 (1100)	300	-	-

Material	Compressive strength. MPa	Flexural strength. MPa	Thermal conductivity. $Wm^{-1}K^{-1}$	Specific heat. $Jkg^{-1}K^{-1}$	Coefficient of thermal expansion $\times 10^{-6o}C^{-1}$	Electrical resistively. μΩcm	Melting or decomposition temperature/ maximum operating temperature. °C
$TiSi_2$	1180	210	-	480	10.4	17	1500/1300
$ZrSi_2$	-	-	-	-	8.3	76	1600/1150
$HfSi_2$	-	-	-	-	-	-	1700/1100
VSi_2	-	-	25	-	11.2	66	1700/1000
$NbSi_2$	-	-	-	-	8.5	50	2050/900
$TaSi_2$	-	-	-	360	9.5	46	2300/800
Cr_3Si_2	-	380	-	-	5.9	80	1560/1150
$MoSi_2$	1130	380	49	550	8.3	20-30	2020/1700
WSi_2	1270	290	45	330	8.2	12	2110/1200

Table 1.6. Some of physical and mechanical properties of nitride ceramics

Material	Crystalline structure	Density. kgm-3	Knoop microhardness. GPa (kg.mm-2)	Young's modulus. GPa	Poisson's ratio	Tensile strength. MPa	Compressive strength. MPa
Covalent:							
β-BN	Cubic	3480	34.3-46.6* (3500-4750)	660	0.15	65	300
Si_3N_4	Hexagonal	3180	19.6 (2000)	310	0.22	350-425	270
AlN	Hexagonal	3260	12.0 (1225)	280	0.25		-
Metallic:							
TiN	Cubic	5440	16.7 (1700)	600	-		-
ZrN	Cubic	7350	14.7 (1500)	500	0.25		-
HfN	Cubic	13940	15.7 (1600)	-	0.25		-
VN	Cubic	6080	13.7 (1400)	460	-		-
Material	Crystalline structure	Density. kgm-3	Knoop microhardness. GPa (kg.mm-2)	Young's modulus. GPa	Poisson's ratio	Tensile strength. MPa	Compressive strength. MPa
NbN	Cubic	8360	13.7 (1400)	480	-		-
TaN	Hexagonal	16360	10.3 (1050)	590	-		-
CrN	Cubic	6140	10.8 (1100)	400	-		-
Mo_2N	Cubic	8040	6.4 (650)	-	-		-
W_2N	Cubic	1220	-	-	-		-

Material	Flexural strength. MPa	Fracture toughness (KIC). MPam1/2	Thermal conductivity. Wm-1K-1	Specific heat. Jkg-1K-1	Coefficient of thermal expansion ×10-6oC-1	Electrical resistively. μΩcm	Melting or decomposition temperature/ maximum operating temperature. oC
Covalent:							
β-BN	110	-	300-600	670	5.0-9.0	1018	3250/1100
Si_3N_4	800-900	3.5-4.0	31	720	3.0	1018	1900/1500
AlN	950	-	36	710	5.8	1017	2200/800
Metallic:							
TiN	-	-	22	595	8.1	20-30	2950/500
ZrN	-	-	10	400	5.9	21	2980/500
HfN	-	-	17	230	6.5	32	3300/500
VN	-	-	5.2	630	8.1	85	2170/450
NbN	-	-	4.3	420	3.2	58	2200/450
TaN	-	-	7.8	190	3.6	198	3090/750
CrN	-	-	12	710	0.7	640	1500/low
Mo_2N	-	-	17	290	1.8	20	900/very low
W_2N	-	-	-	-	-	-	840/very low

* Higher values of hardness are for single-crystal materials.

Table 1.7. Some of physical and mechanical properties of boride ceramics

Material	Crystalline structure	Density. kgm^{-3}	Knoop microhardness. GPa ($kg.mm^{-2}$)	Young's modulus. GPa	Poisson's ratio	Tensile strength. MPa	Compressive strength. MPa
Covalent:							
B	Tetragonal	2340	25.5 (2600)	440	-	2400	-
SiB_6	-	2430	22.6 (2300)	330	-	-	-
AlB_{12}	Orthorhombic/ Tetragonal	2600	25.5 (2600)	430	-	-	-
Metallic:							
TiB_2	Hexagonal	4520	29.4 (3000)	410	-	-	1350
ZrB_2	Hexagonal	6170	22.6 (2300)	340	-	200	1590
HfB_2	Hexagonal	11200	23.5 (2400)	-	-	-	-
VB_2	Hexagonal	5100	27.5 (2800)	260	-	-	-
NbB_2	Hexagonal	7210	24.5 (2500)	650	-	-	-
TaB_2	Hexagonal	12600	19.6 (2000)	250	-	-	-
CrB_2	Hexagonal	5600	21.1 (2150)	540	-	-	1280
Mo_2B_5	Rhombohedral/ Hexagonal	7480	23.5 (2400)	685	-	-	-
W_2B_5	Hexagonal	13100	24.5 (2500)	790	-	-	-

Material	Flexural strength. MPa	Thermal conductivity. $Wm^{-1}K^{-1}$	Specific heat. $Jkg^{-1}K^{-1}$	Coefficient of thermal expansion $\times 10^{-6o}C^{-1}$	Electrical resistively. μΩcm	Melting or decomposition temperature/ maximum operating temperature. °C
Covalent:						
B	530	-	1280	1.1	7×10^{11}	2300/750
SiB_6	-	-	-	5.4	107	1900/-
AlB_{12}	-	-	-	-	2×10^{12}	2210/-
Metallic:						
Material	Flexural strength. MPa	Thermal conductivity. Wm-1K-1	Specific heat. Jkg-1K-1	Coefficient of thermal expansion ×10-6oC-1	Electrical resistively. μΩcm	Melting or decomposition temperature/ maximum operating temperature. oC
TiB_2	250	65	960	4.6	10-30	2790/1400
ZrB_2	100	58	-	5.9	10-15	3245/1300
HfB_2	-	51	300	7.6	12	3250/1500
VB_2	-	42	670	7.6	13	2430/600
NbB_2	-	17	420	8.6	12	3000/850
TaB_2	-	16	250	8.5	14	3090/850
CrB_2	620	32	690	11.1	18	2150/1000
Mo_2B_5	350	27	530	5.0	18	2140/1000
W_2B_5	-	52	460	5.0	18	2365/1000

Table 1.8. Some of physical and mechanical properties of carbide ceramics

Material	Crystalline structure	Density. kgm^{-3}	Knoop microhardness. GPa ($kg.mm^{-2}$)	Young's modulus. GPa	Poisson's ratio	Tensile strength. MPa	Compressive strength. MPa
Covalent							
Diamond	Cubic	3515	78.4-102.0* (8000-10400)	900-1050	0.20	-	-
B_4C	Rhombohedral	2520	31.4 (3200)	400	0.19	300	2850
SiC	β-Cubic, α-hexagonal at 2000°C	3210	23.5-26.5* (2400-2700)	440	0.17	175	1400-3400
Metallic							
TiC	Cubic	4920	27.5 (2800)	450	0.19	250-475	1375-2950
ZrC	Cubic	6560	25.5 (2600)	410	0.19	>200	1650
HfC	Cubic	12670	26.5 (2700)	410	0.18	-	-
VC	Cubic	5480	26.5 (2700)	420	0.22	-	620
NbC	Cubic	7820	22.6 (2300)	450	0.22	250	2400
Material	Crystalline structure	Density. kgm^{-3}	Knoop microhardness. GPa ($kg.mm^{-2}$)	Young's modulus. GPa	Poisson's ratio	Tensile strength. MPa	Compressive strength. MPa
TaC	Cubic	14500	21.6 (2200)	510	0.24	300	-
Cr_3C_2	Orthorhombic	6680	17.7 (1800)	385	0.22	-	1050
Cr_7C_3	Hexagonal	-	-	-	-	-	-
$Cr_{23}C_6$	Cubic	-	-	-	-	-	-
Mo_2C	Hexagonal	9120	16.7 (1700)	530	-	-	900
WC	Hexagonal	15800	20.6 (2100)	690	0.19	350	3500
W_2C	Hexagonal	17150	-	-	-	-	-
WC-Co (94-6%)	Hexagonal	15100	14.7 (1500)	640	0.26	-	3500
high speed steel:M-2 (for refrence)	-	7720	8.5-9.5 (860-960)	200	0.3	-	-

Table 1.8. Some of physical and mechanical properties of carbide ceramics (Continued)

Material	Flexural strength. MPa	Fracture toughness (K_{IC}). $MPam^{1/2}$	Thermal conductivity. $Wm^{-1}K^{-1}$	Specific heat. $Jkg^{-1}K^{-1}$	Coefficient of thermal expansion $\times 10^{-6o}C^{-1}$	Electrical resistively. μΩcm	Melting or decomposition temperature/ maximum operating temperature. °C
Covalent							
Diamond	1000-1050	-	900-2100	530	1.0	10^{13}-10^{28}	3800/1000
B_4C	350	-	26	920	4.3	5×10^5	2420/1100
SiC	675	4.6	85-120	710	4.3	10^5	2830/1700
Metallic							
TiC	860	4	27	520	7.2	60	3065/700
ZrC	250	-	22	250	6.3	42	3440/600
HfC	240	-	13	190	6.3	40-65	3930/600
VC	-	-	10	530	6.7	59	2730/600
NbC	300	-	14	290	7.4	19	3500/650
TaC	300	-	23	190	6.7	35	3900/650
Cr_3C_2	325	-	19	530	9.9	75	1895/1100
Cr_7C_3	-	-	-	-	-	-	-
$Cr_{23}C_6$	-	-	-	-	-	-	-
Mo_2C	50	-	22	315	6.7	57	2500/500
WC	700	8	39	200	5.2	70	2780/500
W_2C	-	-	-	-	-	-	2860/-
WCCo (94-6%)	700-2000	-	90	210	5.4	-	-
high speed steel:M-2 (for refrence)	-	-	28	-	10.6	-	-

* Higher values of hardness are for single-crystal materials.

- The low rate of pollution, compared with other methods
- Disadvantages of composite plating are:
- Due to a relatively low volume of confined particles, electrochemical composite coatings has a lower hardness and strength against abrasives, compared with other dispersion coatings
- A decrease in flexibility feature by adding ceramic particles
- Reducing the adhesion between coating and sub-layer (this problem can be solved through creating a middle non-composite layer)
- Particles agglomeration due to electrolyte inappropriate situation

The following table offers a comparison for a better assessment of metallic matrix composite (MMC) coats obtained from composite plating and other methods.

Table 1.9. Comparison among three important properties in composite plating

particles	Ability to control geometric shape	Suspension ability	Chemical stability
Al_2O_3	○	○	○
B_4C	○	○	◒
BN	◒	◒	○
Cr_3C_2	○	○	◒
Cr_2O_3	○	○	●
Graphite	◒	○	○
Mo	○	◒	●
Mo_2C	○	○	●
MoS_2	◒	◒	●
SiC	○	○	○
Si_3N_4	○	○	○
TiC	○	○	●
TiO_2	○	○	○
WC	○	◒	○
ZrO_2	○	○	○

○Appropriate ◒ approximately appropriate ● Not appropriate.

1.3.3. Dispersive Particles

The particles used in simultaneous electrodeposition process are generally included:

1. high temperature metals of transition groups of periodic table

A. Group IV_b : Ti, Zr, Hf

B. Group V_b: V, Nb, Ta

C. Group VI_b: Cr, Mo, W

2. Ceramics

A. metallic ceramics:

Carbides, nitrites, borides, silicates, and solid solutions

B. non-metallic (ionic) oxides:

Al_2O_3, Cr_2O_3, ZrO_2, HfO_2, TiO_2, SiO_2, BeO, MgO, ThO_2

C. non-metallic (covalent) non-oxides:

B_4C, SiC, BN, Si_3N_4, AlN, Diamond

3. Minerals

TiH_2, CdS, ZnS, $SrSO_4$, $BaSO_4$, CaF_2, WS_2, MoS_2, WSi_2, Fluoride, Glass, Graphite, Kaoline …

4. Organic materials:

Polytetrafluoroethylene (PTFE), Polytyrene, Polypropylene, Polycarbonate, Polyamide, Polyacrylonitril, Polypyrrole, Polyaniline, Acetyl cellulose, Polyvinyl acetate, polyvinyl butyral, Copolymer (methyl methacrylate), Saccharin, Fluorinated carbons (CFx), Resin

Comparing some of physical and mechanical properties of ceramic particles is shown in tables 1.4 to 1.8. Table 1.9 make a comparison among particles required properties for composite plating.

1.4. Simultaneous Deposition Mechanisms

There are many researches about the way of simultaneous deposition of suspending particles with metallic ions. However, for occurrence of many factors which have influence on electrochemical composite deposition, a theory capable of predicting composite chemical composition (volumetric fragments of particles within the coating) and coating structure, has not been introduced yet.

These systems have different variables, including:

- Material, size, shape, and concentration of existed particles in the electrolyte
- Electrolyte's chemical composition (surfactant and additives)
- Electrolyte's temperature
- Current density and type of current (direct or pulse)
- Conditions of mass transfer (type and rate of agitation)
- Electrolyte's pH

1.4.1. Initial Models of Simultaneous Deposition Processes

Initial models offered three possible mechanisms including:

A) Mechanical entrapment

In this state particles move toward cathode due to intense agitation of the electrolyte. If the connection time and metal deposition rate is sufficient, the particles will join to developing metallic layer.

B) Electrophoresis

Charged particles move through the applied electrical field. The electrophoresis speed is directly related to zeta (ζ) potential, which is defined as the potential on shear surface. The shear surface position is somehow conventional. In electrolytes with high concentration, the middle layer is compressed due to high ionic consolidation and consequently ζ tends to be zero. Then it seems that electrophoresis does not have an important role in most of practical cases.

C) Adsorption

Particles are subjected to variant adsorptive forces around the cathode. Particles joining to developing layer might be a physical, chemical, or electrochemical adsorption, which depends on the adsorptive agent type. The particles adsorption phenomenon, an irreversible phenomenon which is addressed as deposition, is physical. Due to more morphological variation and particles shape, this phenomenon is more complex than chemical adsorption of the molecules, which is an irreversible phenomenon named as adhesion.

1.4.2. Guglielmi Classical Model

One of early models, seen in many of articles, is offered by Guglielmi. This model suggests that particles absorbed charge, ζ, is the main factor in determining feasibility of simultaneous deposition. The model is based on two successive adsorption stages. The first stage is the transition of a layer from weak absorbed particles (with relatively high coverage range compared with Helmoltz double layer), which occurs due to intense agitation. At the second stage, first charged particles are transferred at high potential gradient through the electrophoresis adsorption properties on cathode surface, and then particles are absorbed on cathode surface due to coulomb's force between particle and absorbed anions – which causes particles strongly to be absorbed on electrode.

Langmir's offered the following simultaneous deposition model, based on isothermal adsorption theory and Faraday's rules:

$$\frac{\alpha}{1-\alpha} = \frac{nFdv_0}{Wi_0} e^{(A-B)\eta} \frac{kC}{1+kC}$$

where:

α = absorbed particles volumetric ratio
W = atomic mass of deposited metal
d = volumetric mass of deposited metal
F= Faraday number
i_0 = exchanged current
η = over-potential
K = a constant related to required energy value for initiating rate
C = suspended particles concentration
A, B, and υ_0 = deposition constants

Guglielmi offered α versus C/α curve and mentioned that they have a linear relationship in a given potential. The straight line gradient is:

$$tg\phi = \frac{Wi_0}{nFdv_0} e^{(A-B)\eta}$$

When we apply a high voltage in electrolytes, the surface potential is described by Tafel equation ($i/i_0 \approx e^{A\eta}$). In this case equation 2 changes into:

$$\log tg\phi = \log \frac{Wi_0^{B/A}}{nFdv_0} + (1 - \frac{B}{A}) \log i$$

here W, i_0, n, F, A, and B can be defined by metal deposition process. B and K are deposition factors. For instance for SiC particles, 1/K equals to 0.12 and B = 1.51 A. the ratio of accommodated particles in coating is related to value, which in turn depends on reduction rate of absorbed ions on static particles. The model's defect is that the effects of hydrodynamic forces are not taken into account.

1.4.3. Celis Model

There is a more advanced model, with five steps:

1. Ionic cloud development around a particle
2. Particle movement through transition force toward hydrodynamic boundary layer.
3. Particle diffusion within transition layer
4. Particle absorption on cathode surface in presence of ionic cloud
5. Reduction of some ions of ionic cloud as particles reach to metallic matrix

This model is based on this assumption that the only time for a particle to be accommodated in the coating is when a determined fragment of ionic clouds ions are reduced. Since most materials be charged in contact with an aquatic (polar) environment (figure 1.3), first stage does not seem necessary. In this state, anions such as OH^- surround particle surface and develop a dual electric layer.

At the second stage, particle transfer to cathode surface might be done through mechanisms such as displacement, penetration, or electrophoresis, which are basically rely on particle sizes and the applied forces on particles. The applied forces on the particle can be divided into hydrodynamic (resulted from liquids movement through electrolyte agitation) forces, and forces such as gravity, buoyancy, and interactive forces between particle and electrode. For particles of nanometer size, the mechanism is the connective diffusion; while in particles of larger sizes – in micrometer ranges – mobility and gravity forces are more important. For electrical migration, the electrical charge on particles' surface or zeta (ζ) potential seems significant. Adding positive ions such as Tl^+, Cs^+, and NH_4^+, (called cathionic additives or surface active agents) we can raise particle's positive ion.

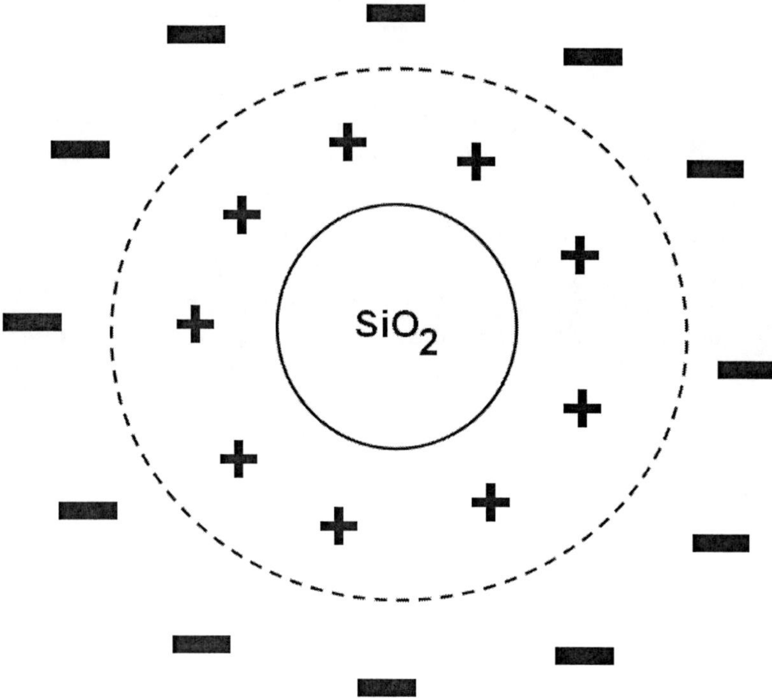

Figure 1.3. Schematic figure of dual layer between the surface of ceramic particle and solution.

1.4.4. Charged Particles Movement in Solutions

Charged particles' electro-kinetic in liquid environment is described by Hükel equation. This equation is obtained from assuming that electrical field attraction forces against particle, equals to solution resistance against movement – according Stoke rule. Particles movement capability is obtained from:

$$U_E = \zeta\varepsilon / (1.5\eta)$$

where, ε and η are solution diffusion ratio and solution viscosity, respectively. Here zeta is obtained from the following equation:

$$\zeta = Q_E / (4\pi\varepsilon a)$$

Raising the particles' electrical charge (Q_E), and decreasing particles' radius (a), zeta potential increases and based on Hükel equation, particles' mobility increases. On the other hand increasing the solution viscosity and decreasing the diffusion, due to high concentration of solution ions, particles' mobility will decline. Here, where the mass transfer of insolvable particles toward cathode get slower, there must be a higher concentration in electrolyte particles to enhance particles' persistence in the coat.

1.4.5. Diffusion Model

Particles' sticking to developing layer leads to a lower concentration of particles on cathode surfaces, and there would be a concentration gradient along the layer. In this stage, as long as the development rate is lower than particles stop time on the surface, they would make a distance from the surface before joining to developing layer and if the development rate is more rapid than stop time, the particles will be accommodated in developing layer. Then, the probability of particles entrapment in the coating is somehow rest upon development rate and, in consequence, deposition current density. Considering Fake's first rule in estimating particles flow along diffusion layer and its relationship with volumetric fraction (x_v) of entrapped particles, we can obtain x_v using the offered model for states a and b - through following equations:

State a:

$$\frac{x_v}{(1-x_v)} = \frac{4\pi r^3 zFN_A}{3V_{m.M}} 1.554 D_p^{2/3} v^{-1/6} \left[c_{p.b} - c^*_{p.b} + c^*_{p.b} \frac{i}{i_{p=1}} \right] \frac{w^{1/2}}{i}$$

State b:

$$\frac{x_v}{(1-x_v)} = \frac{4\pi r^3 zFN_A}{3V_{m.M}} 1.554 D_p^{2/3} v^{-1/6} c_{p.b} \frac{w^{1/2}}{i}$$

According to these equations these equations is proportioned to metallic ion charge (z), particle radius (r), particles' diffusion ratio in the solution (D_p), electrode rotation (w), particle critical concentration in the solution ($c^*_{p,b}$) when development rate is zero, and particles concentration in solution ($c_{p,b}$). Also, it has negatively correlated to metal volumetric molarity of deposited metal ($V_{m,M}$), kinematic viscosity of solution (υ), and deposition current density (i).

1.5. Electrolyte Agitation Effect on Composite Plating

Solution agitation is one of main parameters on plating, especially on composite one. Solution agitation in traditional plating leads to electrolyte unifying in terms of ions distribution and solution temperature. Mechanical agitation of the solution with magnetic agitator and agitating with air or solution persistence circulation by a pump, are variant methods of solution agitation. Effects of solution agitating on composite plating are summed in two items:

1. Creating a homogenous environment of particles – before plating
2. Increasing in simultaneous deposition ratio in electrochemical composite coatings

About the latter, it is said that solution agitating first accelerates ions mobility in electrolyte toward cathode. Also, an increase in agitation speed, somehow because of collision frequency, leads to amount of particles in the coating and in the speeds higher than that - due to segregation in particles with weak adhesion to cathode surface – particles' persistency in the coating have an increase. For instance, the ratio of Silicon Carbide in nickel coating in equal coating condition changes with fluctuation rate changes of electrolyte; as particles volumetric ratio in created coating, obtained from a electrolyte with fluctuation of 50 l/h (7.4%) is higher than the coatings obtained from electrolyte of no oscillation (0.5 %) and electrolytes with oscillation of 500 l/h (6.7%).

1.6. Electrolyte Temperature Effect on Simultaneous Deposition

Increasing electrolyte temperature due to an increase in ions kinetic leads particles more activities in electrolyte, which is helpful for particles simultaneous deposition. On the other hand, according to Langmuir theory raising the temperature causes a decrease in particles adsorption capability, then declining of cathode over-potential and electrical field conducts in a decrease in the ratio of absorbed particles in the coating. Figure 1.4 illustrates the effect of Ni-Co coating of the electrolyte on simultaneous deposition of SiC particles.

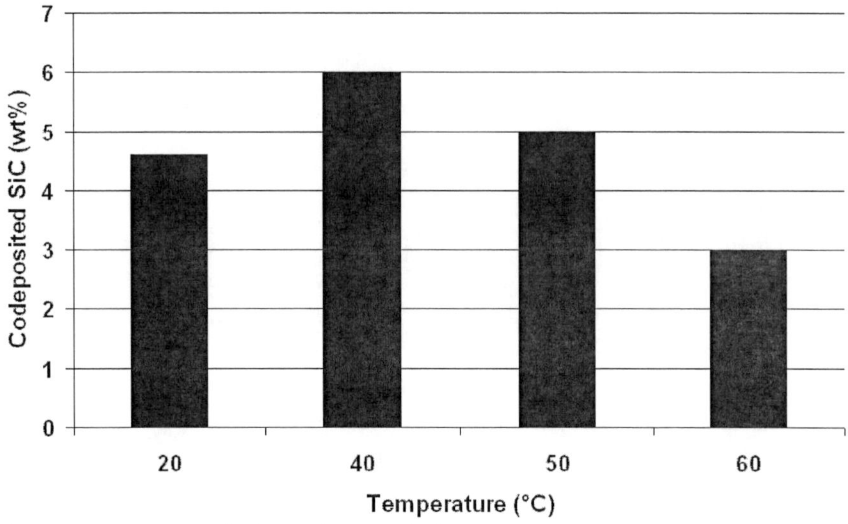

Figure 1.4. weight percentage of SiC particles in Ni-Co coating.

1.7. Electrolyte pH Effect On Simultaneous Deposition

Electrolyte pH has a key role on some metals plating. It is typically seen that in nickel plating, increasing the pH up to 4 causes a raise in rate of absorbed particles on the coat (figure 1.5).

Two reasons have been brought for this:

1. Increasing the pH, amount of H2 molecules on cathode surface decreases and more particles are absorbed on cathode surface.
2. Proton (H+) on ceramic particles' surface prevents form other cations absorption and leads to a weak adsorption of this particles on metallic matrix.

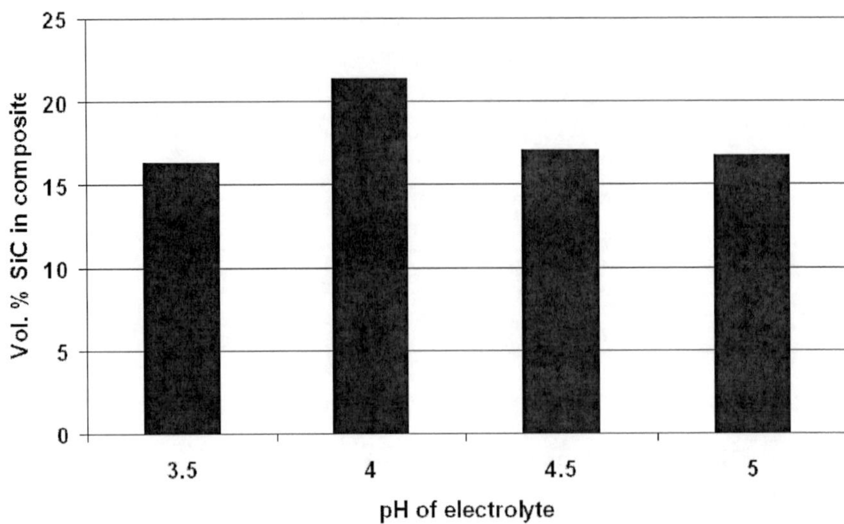

Figure 1.5. Effect of pH on volumetric percentage of SiC particles in composite coating.

1.8. Colloidal Stability

In colloidal systems, where the dispersive phase or colloid has at least one dimension in nanometer to micrometer ranges, colloidal stability or dispersion is considered as an important issue. Generally, the random collisions between dispersive particles take place in a liquid environment. Since composite plating electrolytes are examples of colloidal systems, there is even a chance of insolvable particles adhesion. When these particles stick to electrolytes, an increase in the number of adhered particles in composite's coatings will happen. Therefore, the unique attribution of composite coatings - particularly in the presence of nanoparticles which stick easily to each other due to a high surface energy – will be removed and is considered as an unfavorable factor. Stability of dispersion or particles inclining to agglomeration is defined through particles interaction during the collision. This interaction includes attraction of van der Waals forces towards electrostatic repellent forces, is resulted from overlapping of particles' dual layer. According to Hamaker equation, the van der Waals attraction energy between two spherical particles with radius of a, and intra-particle distance of H (where H>10 nm and H<<a), equals to:

$$V_A = - A\,a / (12\,H)$$

here, A is Hamaker constant an equals to 10^{-19}-10^{-20} J and depends on particles type and their environment. This equation says that van der Waals attraction force increase with an increase in particles radius and their interval distance. On the other hand, the electrostatic repellent energy resulted from overlap dual layers around two spherical particles with equivalent radius of (a) obtained from following equation:

$$V_R = 2 \pi \varepsilon a \psi_d^2 \exp [-k H]$$

where:

ε = permeability coefficient
ψ_d= Stern potential
H = the shortest interval between two particle
In this equation k equals to:
$k = [2 F^2 c z^2/(\varepsilon RT)]^{1/2}$

where:

F = Faraday constant
c = electrolyte concentration
Z = electrolyte ionic capacity
R = ideal gases constant
T = electrolyte temperature (in Kelvin)

According equations corresponding with V_R and k, an increase in particles radius leads to an increase in Stern potential, electrolyte temperature, and also electrostatic repellent force of electrolyte. Overall interaction energy between particles in an electrolyte is obtained from V_R and V_A summation. For avoiding from particles agglomeration in electrolytes and also in composite coating, repellent force among the particles must be greater than their attraction force.

1.8.1. Colloidal Stability Equation with Electrolyte High Concentration

For applying high density current in plating there should be a high concentration of electrode's metallic ions. However, as an increase in electrolyte concentration cause a decrease in diffusion layer thickness, repellent force increases. Then, higher ionic concentration leads to a decline in colloidal stability.

1.8.2. Colloidal Stability Relationship with Zeta Potential

Particles surface charge increase, or in other word Stern potential increase, causes an inclination in repellent force. As potential of zeta shear layer – which is close to Stern layer potential – is measurable, then one can assume the colloidal stability proportioned with zeta potential. Table 1.10 exhibits different amounts of zeta potential in stability states.

1.8.3. The Methods of Colloidal Stability Improvement

Colloidal stability can be improved by two methods including:

1. Adding cationic additives
2. Adding surface surfactants

1.8.3.1. Cationic Additives

Monovalent cationic additives such as Tl+, Cs+, and NH_4+ get easily absorbed by particles and enhance electrostatic repellent forces among particles through modification of particles surface electrical charge. Hence particles agglomeration will take places in the electrolyte with low abundance.

Table 1.10. Zeta potential in stable state

Colloidal stability	Average zeta potential (mV)
Unlimited stability up to good stability	-60 to -100
Good stability	-40 to -60
Medium stability	-30 to -40
Low dispersion	-15 to -30
Near agglomeration	-10 to -15
High agglomeration and deposition	+5 to -5

On one hand these ions attraction, as it previously mentioned, conducts in accelerating of electrical migration in suspended particles of electrolyte. On the other hand simultaneous deposition of particles which are weakly attached to cathode increases on cathode surface – due to their reduction. In this state if cationic additives are the main absorbed ions on particles surface, they might create bounds between particles and matrix phase and even act as spots for electro-crystallization initiation. On the other cases, cationic additives can make a favorable condition for absorbing matrix phase ions on particles surface; since matrix phase ions would be main absorbed category. At this state, there will be an efficient bound between particles and matrix phase and matrix phases' formation will increase. In both states mentioned above cationic additives reduction might strongly prevent growth of matrix phase grains and leads to development of equi-axial grains.

1.8.3.2. Surfactants

Once surfactants exposed to emulsion and suspension solutions and mixtures, they exhibit a great tendency to initiate adsorption interfaces. Thus, it may cause to a decrease in surface tension, wetting, and dispersion. Surfactants are an organic molecule with a part which solvable in water (hydrophilic) and a part which is solvable in lipid (hydrophobic). (figure 1.6)

Regarding surfactants structure they can be classified into four groups: anionic, cationic, non-ionic, and amphoteric. In anionic surfactants the hydrophilic part has negative electrical charge, while the cationic ones have a positive charge. Non-ionic surfactants have no charge and amphoteric ones have both positive and negative charges on each part of molecule. In composite plating mostly cationic and non-ionic surfactants are used.

Some commonly used surfactants in composite plating are:

- Ammonium choloride Cethyl trimethyl (CTAC) which is among most known cationic surfactants.

- Polyelectrolyte dispersant containing aromatic rings (PEDA)
- N-Tetra decyl trimethyl ammonium bromide (TDTAB) which is a cationic surfactant.
- Nonyl phenol ethoxylate (NPE) which is non-ionic surfactant.
- Azobenzene surfactant (AZTAB) which is a cationic surfactant.

Surfactants cause particles dispersion through enhancing their surface electrical potential (electrical charge). Applying surfactants with aromatic cycles (enriched with electron) and with a flat molecular structure, for their easy adsorption and making an organic barrier around particles, we can significantly (or even completely) avoid particles agglomeration.

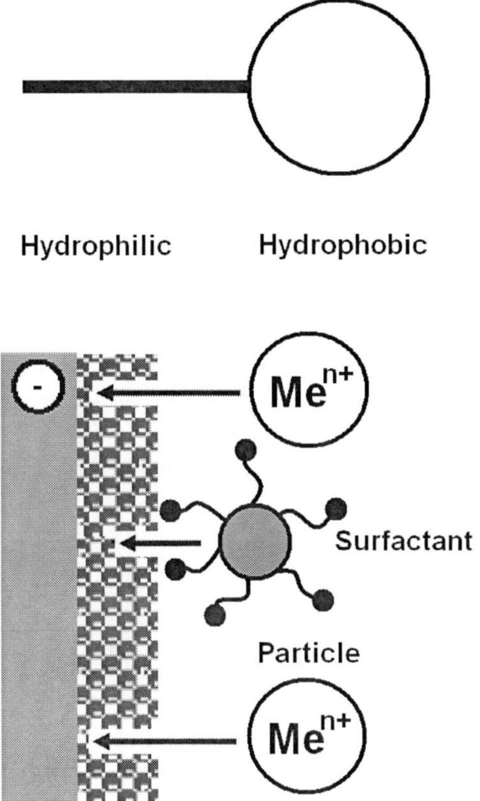

Figure 1.6. (a) Schematic figure of a surfactant (b) performance of a surfactant in an electrolyte.

Table 1.11. Different states in solution transparency with different surfactants concentration

Transparency	*surfactants concentration*
Flocculation	<<
Cloudy solution	<
Milky solution	~
transparent	>

Solution transparency is a determining criterion in colloidal stability and can be identified by naked eye. Depending on surfactants concentration there will be observed different states in solution transparency (table 1.11).

At surfactants with very low concentration the solution seems to be transparence with buoyant particles. In solutions with a little higher concentration, although surfactants are of a little concentration, the solution seems to be blurry or cloudy. As we increase the surfactants ratio, the solution gradually turns into beige colored one with colloidal stability requirement. In surfactants' higher concentrations the solution will be completely transparent.

Surfactants not only are a dispersive agent of solution particles, but also are agent for increasing simultaneous deposition of the particles. In spite of that, there are two difficulties in using common surfactants:

1. The part of surfactants which remains on fine particles surface and is inserted to coating, stops deposition of the other particles. In this state the rate of deposited particles are fairly low.
2. Absorbed surfactants affects coating morphology and might have an unfavorable effects, such as brittleness, on coating.
3. Using Azobenzene modified surfactants, which lose their activity during their simultaneous reduction with metallic ions, these two problems were solved.

A surfactants effect on simultaneous deposition depends on these following factors:

1. Particles dispersing power
2. Surfactants ionic properties
3. Electrodes and particles interaction with surfactants

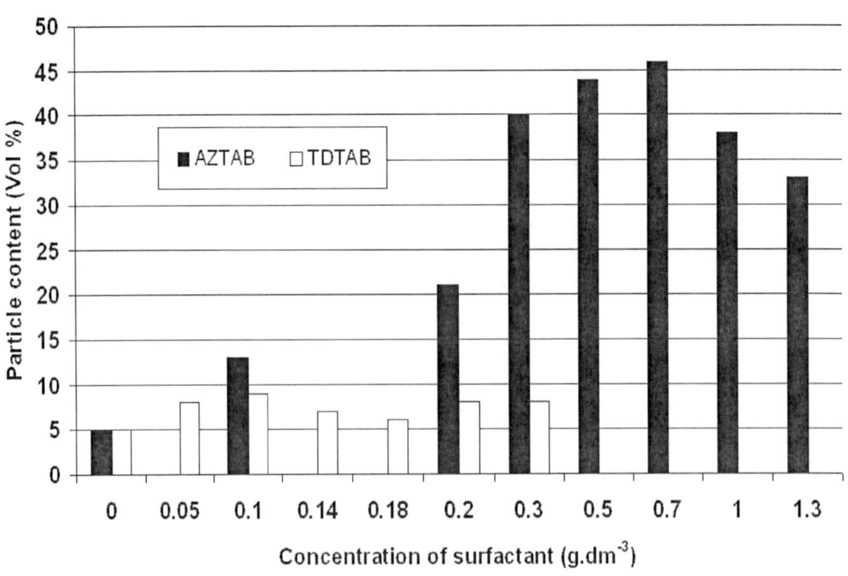

Figure 1.7. Comparison of effects for two surfactants: 1- AZTAB, 2- TDTAB on deposition of B_4C particles with 4 microns in nickel matrix.

Figure 1.7 shows two surfactants effect on deposited particles rate on the coating. In this state both surfactants are cationic and of a same hydrophobicity. Therefore, these surfactants surface activity power of on B_4C particles might be equal. Regarding this one can say the difference between these two surfactants simultaneous deposition is by the difference in their AZTAB molecules electrochemical behaviors compared with TDTAB molecules behavior in electrode-electrode interface. It was observed that AZTAB has a higher reduction potential (-0.15V compared with SCE) than a nickel matrix (-0.48 compared with SCE). Then, AZTAB will be reduced during nickel electrolysis and leads to particles deposition on cathode surface. On the other hand TDTAB is not a reducible surfactant; therefore, absorbed TDTAB molecules on cathode surface can act as a physical barrier against electron transfer or might cause a delay in particles approaching to cathode. Also, the hydrophobic TDTAB layer on cathode surface prevents H_2O molecules and nickel hydrated ions approaching to cathode. All these phenomena in electrode-electrode interface conducts in a decline in particles simultaneous deposition on a nickel matrix.

It was recorded that with 0.75gr SiC, 175mg AZTAB surfactant added to 50ml Watts solution (pH = 1) for nickel plating (where temperature, current density, and time, are respectively 50°C, 10 Adm^{-2}, and 30min), 62.4 volumetric percent of coating is made of SiC.

For creating a dispersive stability in composite plating of the solution, first a defined amount of surfactant is dissolved in the solution. Then, some particles are added to the solution and the mixture is agitated for 5 minutes using ultrasonic waves. Finally, the solution is mechanically agitated before plating for 30min.

1.9. Strengthening Mechanisms

Strengthening mechanisms for electrochemical composite coatings can be summed in two items:

1. Dispersion strengthening using ceramic particles
2. Grain-fining

1.9.1. Dispersion Strengthening Using Ceramic Particles

Displacements in the system composed of disperse particles in a metallic matrix leads to development of displacement cycles, called Orawan. Once coarser particles are distributed separately in the matrix, ceramic particles resistance against movement would be more than that of ceramic. As later movements pass, there will develop a compact mass of displacements, which in turns causes hardness.

Strengthening of edge displacements interaction with non-coherent particles which are not subjected to shearing can be obtained through Orawan equation as following:

$$\Delta\sigma_p = \beta_{OR}\mu bM(f_v)^{1/2}/d_p$$

where:

μ = matrix shear modulus

b = Bergerz vector

f_v = volumetric fraction of particles which are evenly distributed

d_p = particles average size in parallel shear planes

M = Tailor's factor (which is 2.5 to 3 for bcc metals)

$$\beta_{OR} = 1/(2\pi)\ln d_p[\pi/(6f_v)]^{1/2}/b$$

1.9.2. Matrix Grain Fining

Structure sediments fining due to presence of particles - as a peripheral factor for preventing defects development - is one of other strengthening mechanisms of electrochemical composite coatings. In a research, adding aluminum particles to nickel matrix leads to modification of coatings' column structure. This effect is contributed to nickels fine grains initiation on aluminum particles surface; and it is said that in vicinity of aluminum particles and, particularly, coarse grains coatings' growth direction will be changed and the perpendicular line in the coating is not parallel to coating-sublayer interface.

Grains fining prevents displacements and enhance the strength. According to Hall-Petch equation, which is also credible in electrochemical sediments, yield stress has an inverse relationship with metal grain:

$$\sigma_y = \sigma_0 + k\, d^{-1/2}$$

where:

σ_y = yield stress

σ_0 = frictional stress

k = constant coefficient

d = grain diameter

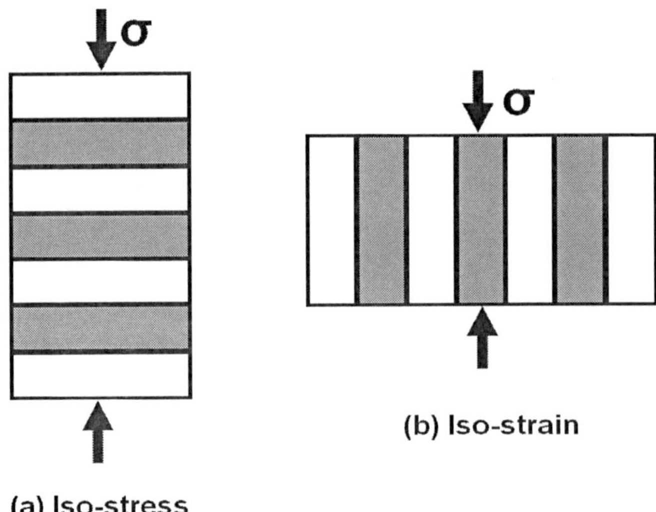

Figure 1.8. Schematic figure for load on different composite structures (a) uni-stress (b) uni-strain.

1.9.3. Effect of Hardness of Dispersed Particles on Coating Hardness

Presence of dispersed particles in electrochemical deposits is similar to reinforces positioning in composite matrix (figure 1.8). Since, proportioned to particles geometric condition of particles a part of them is laid parallel to force exertion direction and other is vertical; however stress's real state in composite is between these two states. Furthermore, as it shown in the figure force exertion direction in hardness measuring is perpendicular to coating's surface. Also, as particles' volumetric ratio in the matrix is low, we can assume a same stress state.

According to rule of mixtures (ROM), composite effective hardness can be obtained by following equation:

$$\bar{H} = (f_m / H_m + f_p / H_p)^{-1}$$

where:

H_m = hardness of matrix
H_p = hardness of particle
f_m = volumetric fraction of matrix
f_p = volumetric fraction of particles

According to this equation an increase in particles hardness results in composite coatings hardness. In this model the effects of particles size and their interspaces are ignored.

1.9.4. Hardness Effect on Strength Against Coating Abrasion

According Archard rule as coating hardness increases, its strength against abrasion raises. The rule is as following:

$$V_w = k\,P\,S / (3\,H)$$

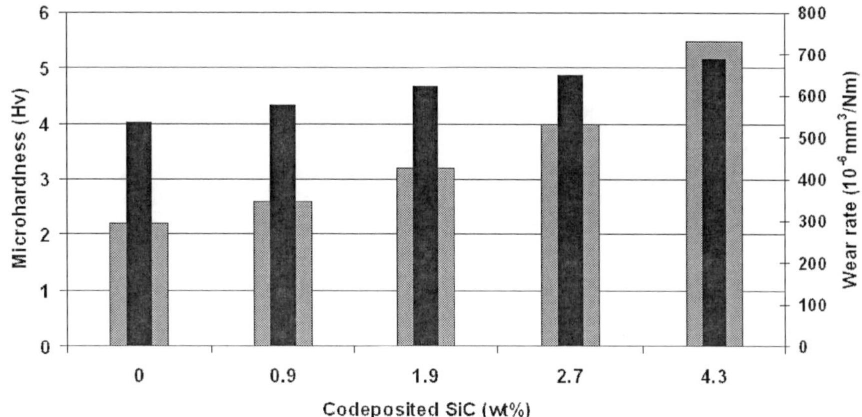

Figure 1.9. Relation among microhardness and wear rate of composite coating reinforced with SiC nanoparticles (average size 50 nm) versus their weight percentage in coating.

where:

V_w = amount of weight loss during abrasion
P = Applied load
S = Shear distance
H = Coating hardness
k = probability of creating abrasive particle due to two surfaces roughness

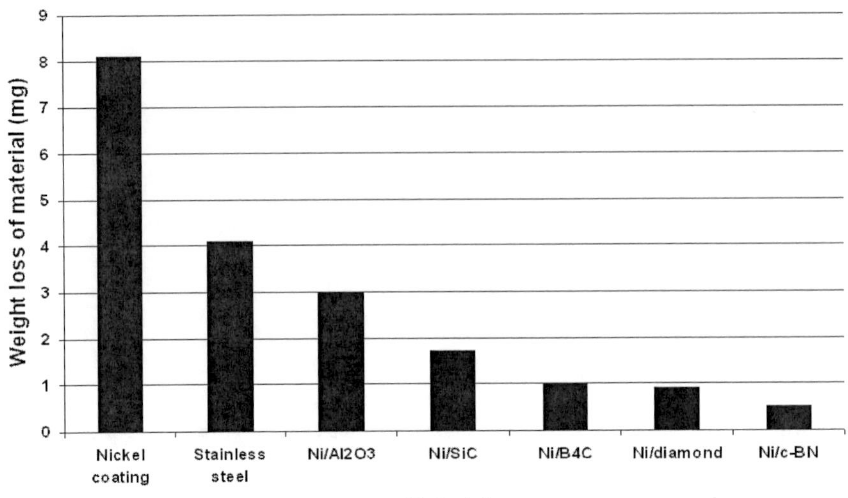

Figure 1.10. Wear resistance of nickel composite coating embedded with different hard ceramic particles with average size of 4 microns.

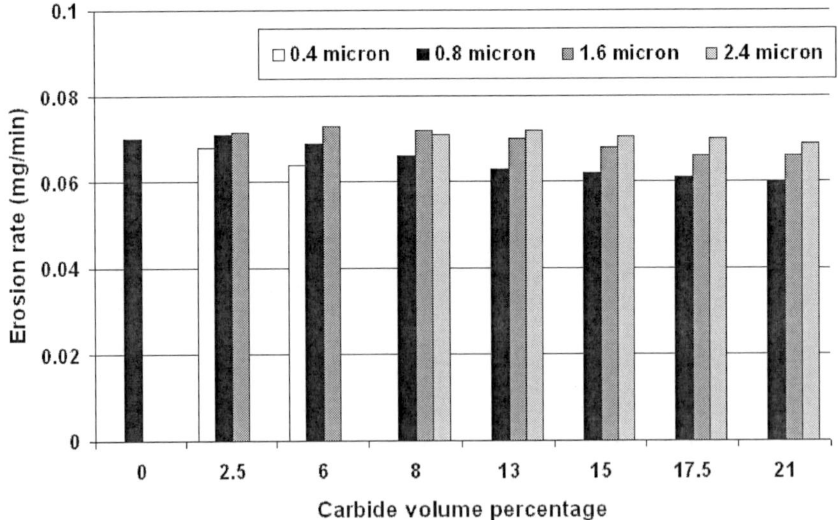

Figure 1.11. Effect of volumetric percentage and size of carbide particles on erosion rate.

Figure 1.9 exhibits the weight ratio effect of accommodated SiC particles in the coating, in terms of hardness and abrasion rate. Regarding introduced items, increasing particles weight ratio, coating's hardness will boost and abrasion rate will decrease.

1.9.5. Particles Effect on Composite Coatings' Mechanical Behavior

Based on offered mechanisms, the mechanical behavior of electrochemical composite coatings depends on following factors:

- Particles' mechanical and physical properties; as the hardness of dispersed particles in the coating is higher, composite coating's strength against abrasion will enhance. Figure 1.10 is an example for this statement. For creating sliding properties, we use PTFE particles, lithium fluoride, calcium carbon, magnesium carbon and fluorinate carbon salts, which each one involves its own advantage. PTFE commonly is used where the applied charge is low, while carbon fluorinate powders are applied on higher temperatures and charges. Fluoride salts lead to lubricating conditions in temperatures close to coatings melting point.
- Volumetric fraction of accommodated particles in the coating; the more the number of particles in the coating, the higher hardness and strength against abrasion
- Particles size; as the particle size decreases hardness and strength of coating against abrasion increases. Figure 1.11 shows carbides volumetric fraction effect (with different sizes) on erosion rate.
- Particles geometric shape: due to a difference in stress field around spherical particles and cubic ones, it seems that they will introduce different mechanical behavior.

1.10. Particles Effect on Coatings Internal Stress

Based on performed studies, it seems that the internal stress of electrochemical composite coatings depends on particles type, amount, and size. For instance, adding organic materials, such as saccharine decreases the nickel coatings' internal stress from Watts' solution, and converts it from tensional to compressive. On the other hand it was said that, once the particles amount is high there will be a strong tensional stress in the coating.

Some researchers believed that micro and macro stresses are caused by a discrepancy between particles mechanical properties and coatings' and recorded that in a nickel coating - in presence of Si_3N_4 particles – for particles with 5.2 μm, as particles number raises, coating's tensional stress will boost too; while for 0.8 μm and 3.6μm particles an increase in particles number leads to a decrease in coatings tensional stress. In this aspect some researches implies that alumina and titanium particles presence, with average of 13 nm and 21 nm diameter, causes a decrease in remained stress in nickel coating.

1.11. Particles Effect on Coatings Corrosion Resistance

There were nontraditional reports about particles effect on composite coatings corrosion resistance. For example, in composite coatings, with nickel-phosphorous matrix, a mixture of phosphides, nickel, and inter-metal particles, results in strong galvanic couples, which speed up corrosion. Hence, it is said that where corrosion resistance is required, nickel electroless composite coatings are not suitable.

On one hand, in coatings with nickel and TiO_2 nano-particle matrix and coatings with nickel and cobalt alloy matrix and SiC nano-particles, corrosion resistance will increase. It is believed that this increase in strength is because of a decrease in nano-composite coatings defects due to corrosive environment and also preventing from corrosive pits development and a speed up in matrix metal deactivation. On the other hand, as it can be seen in table 1.12, corrosion resistance of Zn-Fe coating in presence of SiC has an increase – as corrosion resistance of Zn-Fe-SiO_2 composite coating is 1.5 to 4 times greater than Zn-Fe alloy coating and 3 to 20 times more than pure nickel coating.

1.12. Methods for Determining Coating's Dispersed Particles

1. Gravimetric: in this method first we solve the composite coating in an acidic solution. Then, dispersed particles are separated from the solution - by filtering - and heated. Having composite coating weight (obtained from difference of cathode weight before and after plating) and knowing dispersed particles weight, one can determine particles weight ratio.
2. Microscopic method: this method, also known as image analysis, is commonly used for estimating particles percentage. The cross section of coating is investigated using an optical microscope or a Scanning Electron Microscope (SEM). This method has some errors about fine particles. The method is based on statistical similarity between particles' surface (Fdp) and volumetric (Vdp) concentrations. Therefore, if we determine Fdp by microscopic investigations, then Vdp will be obtained by the following formula:

$$V_{dp} = F_{dp} = Z_{dp} / Z \ \%$$

where:

Z = sum of the special shablon's marks which is matched on microscopic image of the studied surface

Z_{dp} = the number of dispersed marks which are matched on surface

The magnification of 1000 is typically used for this aim. According to the following equation, dispersed particles mass ratio (G_{dp}) is based on V_{dp}, where ρ_{dp} and ρ_M are, respective, the volumetric mass of dispersed particles and matrix.

1. Electronic probe microanalyses: in this method, using scanning electron microscopes' microanalyses (SEM), i.e. WDS (Wavelength dispersive spectroscopy) and EDS (Energy dispersive spectroscopy), and composite coating's element concentration can be defined. Then, comparing this with analysis of dispersed particles standard element, we can calculate particles ratio in the coating.
2. Photon correlation Spectroscopy (PCS).

In this method, the number of dispersed particles in the coating is determined using the rate of reflection of sweep light (produced by Zeta sizer) from particles obtained through composite coating solution in a mixture of Sulfuric and Nitric Acid at temperature of 40 °C. PCS method is a simple and quick method and its relative error is less than 0.2. The method gives satisfactory results even for particles with volumetric ratio of less than 1%, while microscopic and micro-analytical methods introduce repeatable values.

Table 1.12. Comparison among corrosion resistances of Zn, Zn-Fe, Zn-Fe-SiO_2 coatings in 5 wt% sodium chloride solution

Category	Time to appearance Of first corrosion (hr)	Time to appearance of first corrosion (hr)	Time to appearance of 90% corrosion (hr)
Zn	13	166	326
Zn-Fe Fe = 8.86%	66	426	579
Zn-Fe-SiO_2 Fe = 8.86 wt% SiO_2 = 0.47 wt%	256	579	1150

REFERENCES

[1] Aal, A.A., El-Sheikh, S.M. and Ahmed, Y.M.Z. Electrodeposited composite coating of Ni-W-P with nano-sized rod- and spherical-shaped SiC particles. *Materials Research Bulletin*, 2009, 44(1), 151-159.

[2] Abdel Aal, A. and Hassan, H.B. Electrodeposited nanocomposite coatings for fuel cell application. *Journal of Alloys and Compounds*, 2009, 477(1-2), 652-656.

[3] Aliofkhazraei, M., Ahangarani, S. and Sabour Rouhaghdam, A. Effect of the duty cycle of pulsed current on nanocomposite layers formed by pulsed electrodeposition. *Rare Metals,* 2010, 29(2), 209-213.

[4] Aliofkhazraei, M., Ahangarani, S.H. and Rouhaghdam, A.S. Effect of surface nanocrystallization and PPEC time on complex nanocrystalline hard layer fabricated by plasma electrolysis. *Transactions of Nonferrous Metals Society of China*, 2010, 20(3), 425-431.

[5] Aliofkhazraei, M., Hassanzadeh-Tabrizi, S.A., Sabour Rouhaghdam, A. and Heydarzadeh, A. Nanocrystalline ceramic coating on [gamma]-TiAl by bipolar plasma electrolysis (effect of frequency, time and cathodic/anodic duty cycle). *Ceramics International*, 2009, 35(5), 2053-2059.

[6] Aliofkhazraei, M., Morillo, C., Miresmaeili, R. and Sabour Rouhaghdam, A. Carburizing of low-melting-point metals by pulsed nanocrystalline plasma electrolytic carburizing. *Surface and Coatings Technology,* 2008, 202(22-23), 5493-5496.

[7] Aliofkhazraei, M., Rouhaghdam, A.S., Ghobadi, E. and Mohsenian, E. Electrodeposition and mechanical and corrosion resistance properties of tertiary Ni-W/Al_2O_3/CNT nanocomposite coatings. *Advanced Materials Research*, 2010, (89-91), 12-16.

[8] Aliofkhazraei, M. and Sabour Roohaghdam, A. A novel method for preparing aluminum diffusion coating by nanocrystalline plasma electrolysis. *Electrochemistry Communications*, 2007, 9(11), 2686-2691.

[9] Aruna, S.T., Bindu, C.N., Ezhil Selvi, V., William Grips, V.K. and Rajam, K.S. Synthesis and properties of electrodeposited Ni/ceria nanocomposite coatings. *Surface and Coatings Technology*, 2006, 200(24), 6871-6880.

[10] Ataee-Esfahani, H., Vaezi, M.R., Nikzad, L., Yazdani, B. and Sadrnezhaad, S.K. Influence of SiC nanoparticles and saccharin on the structure and properties of electrodeposited Ni-Fe/SiC nanocomposite coatings. *Journal of Alloys and Compounds*, 2009, 484(1-2), 540-544.

[11] Avramova, I., Stefanov, P., Nicolova, D., Stoychev, D. and Marinova, T. Characterization of nanocomposite CeO_2-Al_2O_3 coatings electrodeposited on stainless steel. *Composites Science and Technology*, 2005, 65(11-12), 1663-1667.

[12] Chang, S.T., Leu, I.C. and Hon, M.H. Novel methods for preparing nanocrystalline SnO_2 and Sn/SnO_2 composite by electrodeposition. *Journal of Alloys and Compounds*, 2005, 403(1-2), 335-340.

[13] Chen, J.S., Huang, Y.H., Liu, Z.D. and Tian, Z.J. Jet electrodeposited Cu-Al_2O_3 nanocomposite coatings. *Proceedings of the International Conference on Integration and Commercialization of Micro and Nanosystems* 2007, pp. 1227-1230.

[14] Cheong, M. and Zhitomirsky, I. Electrodeposition of alginic acid and composite films. *Colloids and Surfaces A: Physicochemical and Engineering Aspects*, 2008, 328(1-3), 73-78.

[15] Fustes, J., Gomes, A. and Da Silva Pereira, M.I. Electrodeposition of Zn-TiO_2 nanocomposite films-effect of bath composition. *Journal of Solid State Electrochemistry*, 2008, 12(11), 1435-1443.

[16] Hosseini, M.G., Abdolmaleki, M., Sadjadi, S.A.S., Boroujen, M.R., Arshadi, M.R. and Khoshvaght, H. Electrodeposition of Ni-WndashB nanocomposite from tartrate electrolyte as alternative to chromium plating. *Surface Engineering*, 2009, 25(5), 382-388.

[17] Jeon, Y.S., Byun, J.Y. and Oh, T.S. Electrodeposition and mechanical properties of Ni-carbon nanotube nanocomposite coatings. *Journal of Physics and Chemistry of Solids,* 2008, 69(5-6), 1391-1394.

[18] Khazrayie, M.A. and Aghdam, A.R.S. Si_3N_4/Ni nanocomposite formed by electroplating: Effect of average size of nanoparticulates. *Transactions of Nonferrous Metals Society of China*, 2010, 20(6), 1017-1023.

[19] Lakatos-Varsunyi, M., Mik, A., Varga, L.K. and Kulman, E. Electrodeposited magnetic multi-nano-layers. *Corrosion Science*, 2005, 47(3 SPEC. ISS.), 681-693.

[20] Li, J., Sun, Y., Sun, X. and Qiao, J. Mechanical and corrosion-resistance performance of electrodeposited titania - Nickel nanocomposite coatings. *Surface and Coatings Technology*, 2005, 192(2-3), 331-335.
[21] Liu, Y., Ren, L., Yu, S. and Han, Z. Influence of current density on nano-Al_2O_3/Ni+Co bionic gradient composite coatings by electrodeposition. *Journal of University of Science and Technology Beijing: Mineral Metallurgy Materials* (Eng Ed), 2008, 15(5), 633-637.
[22] Liu, Y., Yu, S.R., Ren, L.Q. and Han, Z.W. nano-Al_2O_3/Ni + Co gradient composite coating by electrodeposition. Jilin Daxue Xuebao (Gongxueban)/*Journal of Jilin University* (Engineering and Technology Edition), 2009, 39(SUPPL. 1), 154-158.
[23] Mirzamohammadi, S., Aliov, M.K., Sabur, A.R. and Hassanzadeh-Tabrizi, A. Study Of Wear Resistance And Nanostructure For Tertiary Al_2O_3/Y_2O_3/CNT Pulsed Electrodeposited Ni-Based Nanocomposite. *Materials Science,* 2010(1).
[24] Mirzamohammadi, S., Kiarasi, R., Aliov, M.K., Sabur, A.R. and Hassanzadeh-Tabrizi, A. Study of corrosion resistance and nanostructure for tertiary Al_2O_3/Y_2O_3/CNT pulsed electrodeposited Ni based nanocomposite. *Transactions of the Institute of Metal Finishing*, 2010, 88(2), 93-99.
[25] Pang, X. and Zhitomirsky, I. Electrodeposition of composite hydroxyapatite-chitosan films. *Materials Chemistry and Physics*, 2005, 94(2-3), 245-251.
[26] Pang, X. and Zhitomirsky, I. Electrodeposition of nanocomposite organic-inorganic coatings for biomedical applications. *International Journal of Nanoscience*, 2005, 4(3), 409-418.
[27] Pang, X. and Zhitomirsky, I. Electrodeposition of hydroxyapatite-silver-chitosan nanocomposite coatings. *Surface and Coatings Technology,* 2008, 202(16), 3815-3821.
[28] Peng, X., Yan, J., Xu, C. and Wang, F. Oxidation at 900 °C of the chromized coatings on A3 carbon steel with the electrodeposition pretreatment of Ni or Ni-CeO_2 film. *Metallurgical and Materials Transactions A: Physical Metallurgy and Materials Science*, 2008, 39(1), 119-129.
[29] Peng, X., Zhao, J., Zhang, H. and Wang, F. Novel electrodeposited Ni-based nanocomposite precursors for nitriding and low temperature chromizing. *Materials Science Forum*, 2006, pp. 331-338.
[30] Ramalingam, S., Muralidharan, V.S. and Subramania, A. Electrodeposition and characterization of Cu-TiO_2 nanocomposite coatings. *Journal of Solid State Electrochemistry*, 2009, 13(11), 1777-1783.
[31] Shi, L., Sun, C., Gao, P., Zhou, F. and Liu, W. Mechanical properties and wear and corrosion resistance of electrodeposited Ni-Co/SiC nanocomposite coating. *Applied Surface Science*, 2006, 252(10), 3591-3599.
[32] Sun, X.J. and Li, J.G. Friction and wear properties of electrodeposited nickel-titania nanocomposite coatings. *Tribology Letters*, 2007, 28(3), 223-228.
[33] Sun, X.J. and Li, J.G. Tribological characterisation of electrodeposited nickel - Titania nanocomposite coatings sliding against silicon nitride in high vacuum. *Surface Engineering*, 2008, 24(3), 236-239.

[34] Thiemig, D. and Bund, A. Characterization of electrodeposited Ni-TiO_2 nanocomposite coatings. *Surface and Coatings Technology*, 2008, 202(13), 2976-2984.

[35] Tsai, Y.C., Li, S.C. and Liao, S.W. Electrodeposition of polypyrrole-multiwalled carbon nanotube-glucose oxidase nanobiocomposite film for the detection of glucose. *Biosensors and Bioelectronics*, 2006, 22(4 SPEC. ISS.), 495-500.

[36] Tu, W.Y., Xu, B.S., Dong, S.Y. and Wang, H.D. Electrocatalytic action of nano-SiO2 with electrodeposited nickel matrix. *Materials Letters*, 2006, 60(9-10), 1247-1250.

[37] Vaezi, M.R., Sadrnezhaad, S.K. and Nikzad, L. Electrodeposition of Ni-SiC nano-composite coatings and evaluation of wear and corrosion resistance and electroplating characteristics. *Colloids and Surfaces A: Physicochemical and Engineering Aspects,* 2008, 315(1-3), 176-182.

[38] Wang, W., Qian, S.Q., Zhou, X.Y. and Lin, W.S. Microstructure and properties of high speed jet electrodeposition nanocomposite coatings. *Cailiao Rechuli Xuebao/Transactions of Materials and Heat Treatment,* 2007, 28(SUPPL.), 243-248.

[39] Wang, W., Qian, S.Q., Zhou, X.Y., Lin, W.S. and Yan, M.J. Microstructure and corrosion properties of Ni/PTFE nanocomposite coatings by high speed jet electrodeposition. *Jinshu Rechuli/Heat Treatment of Metals,* 2006, 31(SUPPL.), 149-152.

[40] Xia, F., Jia, Z., Wu, M. and Li, Z. Effect of surfactants on Ni-TiN nanocomposite coatings prepared by ultrasonic electrodeposition. *Chinese Journal of Mechanical Engineering* (English Edition), 2008, 21(3), 84-86.

[41] Xia, F.f., Wu, M.h., Wang, F., Jia, Z.y. and Wang, A.l. Nanocomposite Ni-TiN coatings prepared by ultrasonic electrodeposition. *Current Applied Physics,* 2009, 9(1), 44-47.

[42] Xue, Y.J., Zhu, D., Jin, G.H. and Zhao, F. Friction and wear properties of electrodeposited Ni-La_2O_3 nanocomposite coatings. *Mocaxue Xuebao/Tribology,* 2005, 25(1), 1-6.

[43] Yang, X.y., Li, K.j., Peng, X. and Wang, F.h. Beneficial effects of Co^{2+} on co-electrodeposited Ni-SiC nanocomposite coating. *Transactions of Nonferrous Metals Society of China* (English Edition), 2009, 19(1), 119-124.

[44] Yao, Y., Yao, S., Zhang, L. and Wang, H. Electrodeposition and mechanical and corrosion resistance properties of Ni-W/SiC nanocomposite coatings. *Materials Letters*, 2007, 61(1), 67-70.

[45] Zhang, W.F., Zhu, D. and Zeng, Y.B. Study of optimizing process parameters about Ni-SiC composite electrodeposition based on BP neural network. *Gongneng Cailiao/Journal of Functional Materials*, 2004, 35(3), 383-384+388.

[46] Zhang, Y., Zhang, C., Peng, X. and Wang, F. Effect of Cr particle size on the oxidation behaviour of electrodeposited Ni-Cr composite coatings. *Journal of the Chinese Society of Corrosion and Protection*, 2006, 26(2), 85-88.

[47] Zhang, Z., Niu, Z.X., Zhang, J.Q. and Cao, C.N. Electrodeposition of Ni-SiC nanocomposite coatings based on the surface charge determination of SiC nanoparticles. *Bulletin of Electrochemistry*, 2006, 22(4), 189-192.

[48] Zhao, G.g., Zhou, Y.b. and Zhang, H.j. Sliding wear behaviors of electrodeposited Ni composite coatings containing micrometer and nanometer Cr particles. *Transactions of Nonferrous Metals Society of China* (English Edition), 2009, 19(2), 319-323.

[49] Zheng, H.Y. and An, M.Z. Electrodeposition of Zn-Ni-Al_2O_3 nanocomposite coatings under ultrasound conditions. *Journal of Alloys and Compounds,* 2008, 459(1-2), 548-552.

[50] Zhou, Y.B., Peng, X. and Wang, F.H. Oxidation of a novel electrodeposited Ni - 28.0 mass% Al nanocoating. *Corrosion Science and Protection Technology,* 2005, 17(4), 219-222.

Chapter 2

INTRODUCTION TO DIFFERENT METHODS FOR FABRICATION OF NANOCOMPOSITE COATINGS

ABSTRACT

This chapter discusses about different methods for fabrication of nanocomposite coatings. It starts by the connection of nanocomposite coatings to nanotechnology and continues by some relevant methods for from surface engineering and goes through different technologies for fabrication of nanocomposite layers. This chapter also discusses some of the properties which will affected by fabrication method. Ceramics and ceramic composites' usage limitation is some of their mechanical properties, such as their brittleness or their relatively low ductility. During past years, some researches prove that developing nano structures improve some mechanical properties in ceramics, such as toughness and strength. Generally, there are different opinions about nano-materials, but their unique properties have been focused by many researchers. For instance, nano-particles might leads to thermal and electrical conductivity in insulators. All in all, constructing composites with nano-structure requires nano rough materials. There are different methods for producing nano-powder, but it must be considered that primary powder type has a real important effect on last piece's properties.

Table 2.1. Composite panels used in a plane

	Cargo floor panel	Cargo floor beam	Air conditioning duct	VOR antenna
Composite part	-Fibber glass / epoxy / aluminum -Balsawood core	- Stitched carbon fiber / epoxy -RTM	- Carbon fiber / BMI	S-2 fibber glass / epoxy

	S-duct	Nacelle	VSCF tunnel scope	Nose strake	VSCF nose scoop
Composite part	- Carbon fiber / epoxy - Nomex honeycomb	- Carbon fiber / epoxy - Aluminum honeycomb	- fiber glass / epoxy - Integral deicing	- Fiber glass / epoxy - Integral deicing	- fiber glass / epoxy - Integral deicing

2.1. INTRODUCTION

A development in technology demands new materials in different industries. There are materials with variant properties used in different industries - based on their application. For instance material used in aerospace should involve high strength and sufficient thermal stability, in addition to their low weight. Hence, a group of material, referred as composites, was developed. These materials have a key role in different industries. Compared with simple materials, they have better thermal and mechanical properties. Table 2.1 shows composite panels used in a plane.

Composites are, typically, composed of two or more components, where one component serves as matrix while the others are reinforcements. Reinforcement phase can be existed as particles or fibers. Composite's matrix can be polymeric, metallic, or ceramic. Although Ceramic composites are efficient in high-temperature applications, but brittleness is their weakness. In spite of that, some researches show that creating nano-structure can improve their mechanical characteristics, particularly their ductility.

A good choice of ceramic materials for producing ceramic composites is of a great importance. Non-oxide ceramics, such as SiC and Si_3N_4, have some significant mechanical properties at high temperatures; however they will oxidize at temperatures over 1500 °C in air proximity. Although oxide ceramics enjoy a better strength against oxidation and can be better sintered than non-oxide ones, but they have a lower creep strength which limits their applications. For example, alumina is one mostly used engineering ceramics with many usages in different industries. It is also applied in high temperatures and shows slight creep properties, being subjected to stresses. For their creep improvement a material with high creep strength from secondary phase will be used- or as one can say an alumina composite will be made. However different materials are used in alumina bearing composites, but among them YAG (Yttrium Aluminum Garnet, $Y_3Al_5O_{12}$) is one of the best choices for enhancing alumina creep strength – as a mono crystal of YAG has ten times greater creep strength than alumina. YAG's particular structure stops movements and displacements.

Ceramics and ceramic composites' usage limitation is some of their mechanical properties, such as their brittleness or their relatively low ductility. During past years, some researches prove that developing nano structures improve some mechanical properties in ceramics, such as toughness and strength. For this, nano-materials have been developed during recent years. For their special properties, nano-materials are used in different industries, such as electronics and new applications. Generally, there are different opinions about nano-materials, but their unique properties have been focused by many researchers. For instance, nano-particles might leads to thermal and electrical conductivity in insulators.

All in all, constructing composites with nano-structure requires nano rough materials (powder). There are different methods for producing nano-powder, but it must be considered that primary powder type has a real important effect on last piece's properties. For example, using two nano-powders separately can lead to aggregation of homogeneous particles along each other. This come from powders high amount of surface energy and their tendency for decreasing their energy. This can be efficiently handled using chemical methods such as sol-gel. Generally, the sol-gel method is a decent way for producing composite nano ceramic based powders with high quality and nano structure. Through this method one can obtain several ceramic oxides in completely homogenous state. This method, also, cause to a decrease in sintering temperature, which is due to a high special surface of fabricated powder.

In some studies, YAG-alumina composite powder was synthesized using hydrated aluminum chloride, aluminum powder, and Yttrium oxide, with sol-gel method. The effect of aluminum and Yttrium on gel making processes and phase changes of composite nano-powder was investigated. Since in sol-gel method usually organic materials or inorganic materials are used for sol stabilizing and also the method demands an accurate control of pH, it was tried to use a modified method without any need for sol stabilizer or pH control. After synthesizing, the obtained composite powder by this method is compressed and sintered in different temperatures.

2.2. Nano-Technology and Its Role on Ceramics

Nano-technology is a concept which is addressed to all advance technologies in the field of work with nano scale. Usually when it is said nano scale, it means 1 to 100 nm. However, it must be added there are no distinct boundaries for this definition, as in some references 1 to 250 nanometers is considered in nano range. But the important thing is that there are some characteristics which are shown in nano scale materials, as it is expected that, materials with nano structures have better mechanical, physical, and biological properties than that of materials with micro structures. In addition, in nano ranges powders can be better sintered.

The first traces of nano-technology - however not known by this name - came back to 1959. In this year Richard Fineman during a lecture - called "there are much spaces in lower levels" - introduced the notion of nano-technology. Through this theory he announced that molecules and atoms can be directly manipulated in the near future. Nanotechnology gradually entered into sciences such as physics, chemistry, medicine, and, particularly, materials science. The technology's appeals in ceramic science can be summarized as following:

Ceramics has some special characteristics such as high thermal resistivity, and chemical stability. Unlike these features they have low ductility which is due to ceramics nature and atomic boundaries type (ionic or covalent). Getting rid of such a problem a new kind of materials - addressed as composites - were developed which are composed of two or more components. Parallel to scientific advancements and introducing nanotechnology, nanocomposites have been of a great interest. The results show these composites have higher ductility, in comparison with microcomposites.

2.3. Nanocomposite Types

Generally one can categorized nano composites into three groups. This classification is based on distributed phases type and apparent shape in matrix. As it can be seen in figure 2.1, there are three types of reinforcing materials, including layered, fibrous, and particulate.

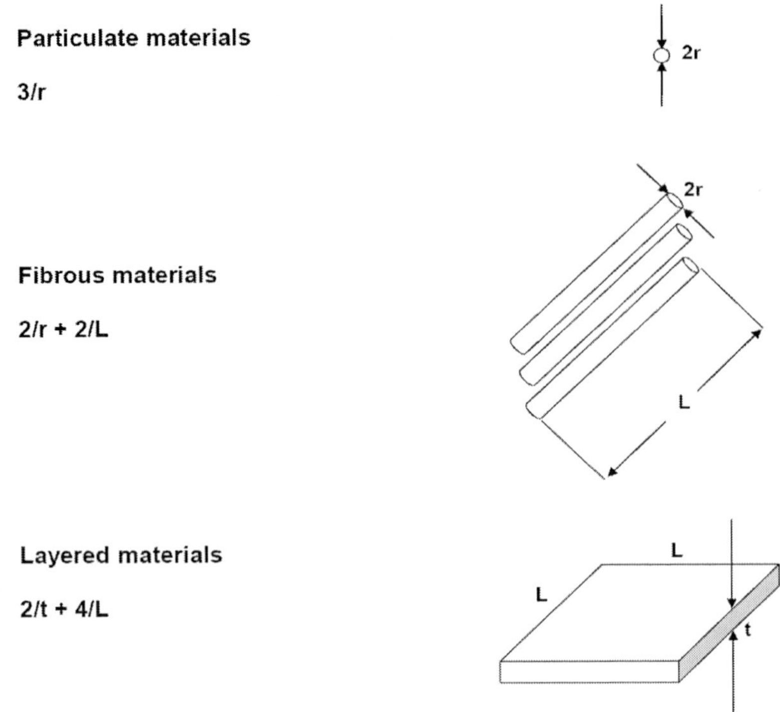

Figure 2.1. Types of composites secondary reinforcing phase.

As these particles size decreases, their surface to volume ratio increases which can introduce new features to composites.

2.3.1. Nano-Platelet Reinforced Composites

In this state nano-platelets are distributed within a matrix. Platelet uniform distribution is an important subject which must be considered. For this group we can name clay/polymer composites. In a typical state clay layers intervals are minimum. Once polymeric matrix diffuses into clay layers, layers intervals will increase. A polymeric matrix, also, can completely separate layers from each other. Each of these states can create different characteristics in composites. Applying clay/polymer nano-composites improves properties such as strength, hardness, ductility, and thermal stabilization. Nowadays, these composites are used in manufacturing many vehicle panels.

2.3.2. Nano-Fiber Reinforced Composites

In these composites fibers are distributed in a matrix. According their usage, the fibers can have different materials. One of materials which are highly interested during these days is carbon nano-tubes. For instance, producing a nano-composite with alumina matrix, researchers provide an alumina sol, using an aluminum Alco-oxide, and then alcohol and carbon nano-tube mixture is added to sol, which turns the sol into a gel. Finally, after calcination process, a composite of alumina-CNT will be obtained. Similar to previous ones, uniform distribution of secondary phase have a great effect on final nano-composites properties. In addition, the interface form between matrix and fibers is of a great importance; as more chemical boundaries (versus to a weak physical boundary) have higher impact on strength and ductility enhancement.

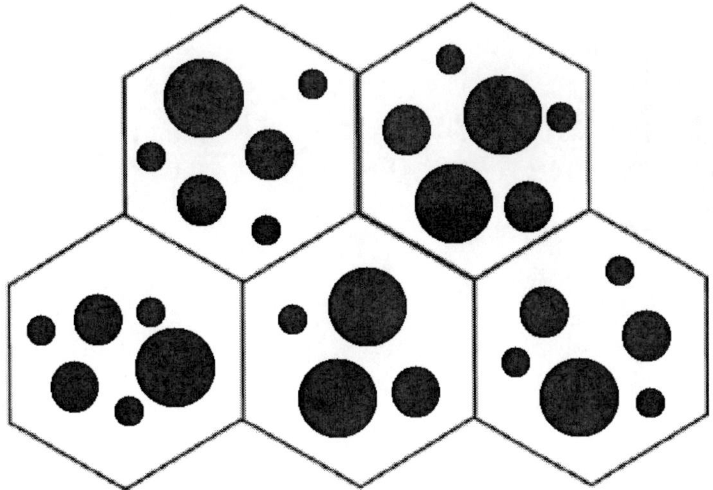

Figure 2.2. Schematic figure of nano-particle reinforced composites.

2.3.3. Nanoparticle Reinforced Composites

In this case particles of secondary phase are distributed in matrix (figure 2.2). May one can say these form of composites are more common than the others. They classified into three categories:

1. Intra-type nanostructure composites
2. Inter-type nanostructure composites
3. Nano/nano nanostructure composites

Some researches carried on nano-composites, suggested that using nano-materials conducts in a significant enhancement of strength and ductility. Also, some other mechanical properties, such as hardness, resistance against abrasion, and resistance against thermal shock are improved.

2.4. Nanocomposites Toughness Model and Mechanism

Based on offered models, in nano-composites there would be remaining stresses around nano-particles distributed within matrix. This is due to a difference between thermal expansion coefficient of matrix and distributed nano phase. The magnitude of this stress quickly declines, as we get far from secondary phase. Since this secondary phase is in nano range, the stress just only causes a displacement around it (figure 2.3). The possibility of occurring big defects in nano-composites is highly rare. Due to anneal process, these displacements turn into minor boundaries around nano phase.

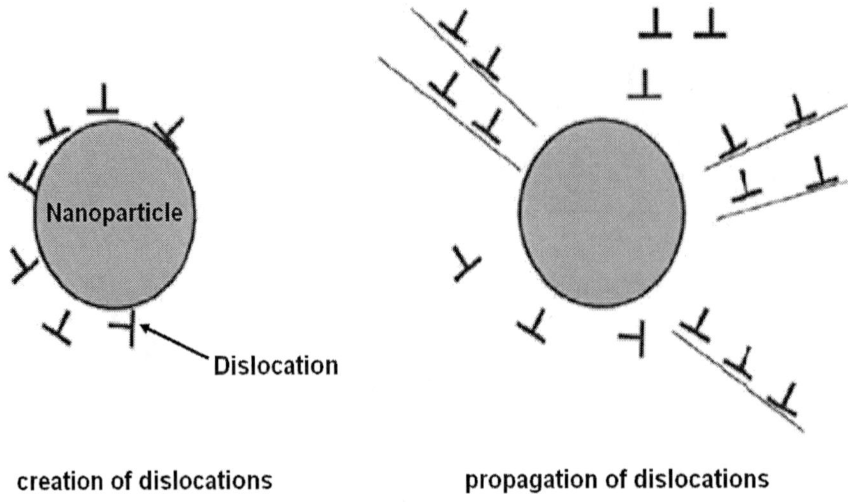

Figure 2.3. Effect of nanoparticles presence within matrix.

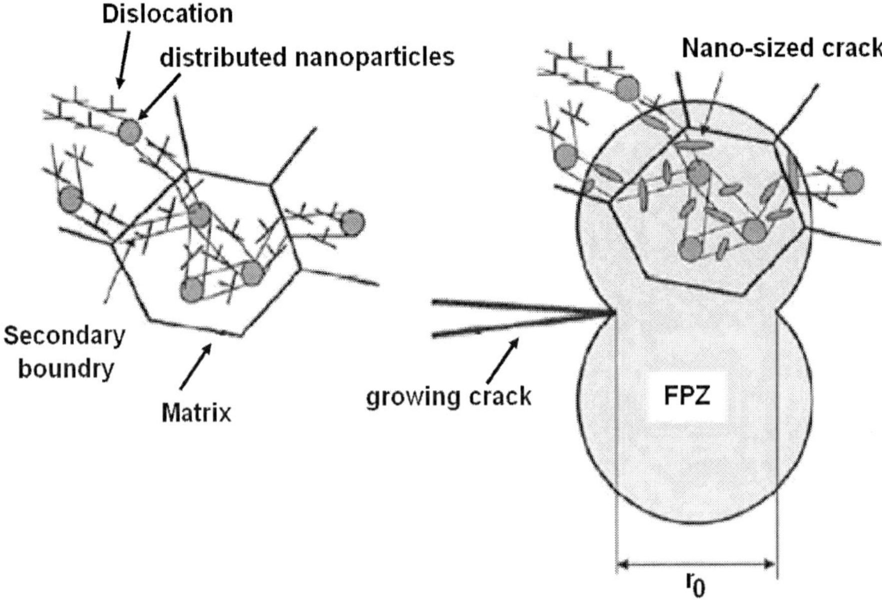

Figure 2.4. Illustration of toughness mechanism in nano-composites.

According the model and considering figure 2.4, one can describe toughness mechanism in nano-composites. In figure 2.4.a, created minor boundaries through displacements during annealing process is shown. In this state as the crack is developing, their reaching to these displacements makes them to serve as a source for creating nano-cracks, which stop crack development through energy adsorption (figure 2.4.b).

2.5. Methods for Fabrication of Nano-Materials

As it previously mentioned, rough materials have an important role in fabrication of composites. For instance, for producing composites with nano-structures one must use nano rough materials; or in the cases of applying powder method for production, primary properties of powders, such as sinterability, uniformity, etc, have a significant effect on panel's final characteristics.

In this section some methods for producing nano-materials will briefly discussed.

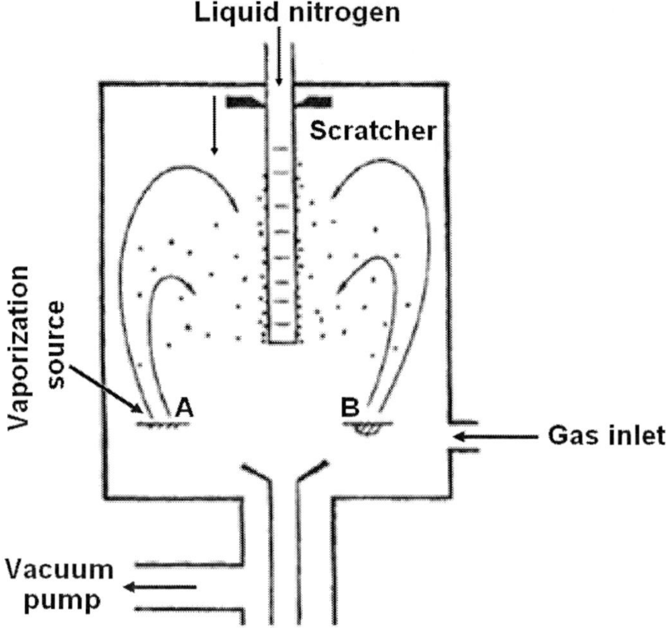

Figure 2.5. Device for producing nano-powder through vapor route.

2.5.1. Method of Producing Nano-Materials from Vapor Route

The method includes vapor physical and chemical deposition. At vapor physical deposition method, the given atoms are picked from the source through the energy and deposited in considered point. In the vapor chemical deposition reagent gases in a particular surface, react with each other until forming the desired material. One of this method's limitations is the device's high cost. Figure 2.5 illustrates one of these devices for fabrication of nano-powder.

2.5.2. Mechanical Methods

In this method the powder is ground through the mechanical force and gets smaller. Some effective parameters in milling are:

1. Applied materials ratio
2. Materials hardness
3. Balls size distribution
4. Mill's rotation speed

Controlling these parameters one can reach the favorable result. However, impurities intrusion into samples, through milling process, makes this method completely limited. On the other hand, obtaining nano-ranges a great deal of energy is required. Figure 2.6 demonstrates abrasion via ball mill.

2.5.3. Precipitation Method

Precipitation method is one of the chemical methods. The method is based on the dual layer developed layer on suspending particles in the sol. Sol particles precipitation is basically due to the following feature:

1. A decrease in surface potential due to pH changes
2. An increase in electrolyte concentration in the solution

A particles having charges is related to its surrounding environment pH. Regarding this, as pH tends to be values with approximately zero surface charges, sol will be more unstable and would precipitate. Figure 2.7 shows alumina stability diagram.

Considering the figure, reaching to pH = 9 causes sol instability and precipitation. This mechanism is the basis of precipitation method. Also, increasing salt concentration, there would be alumina particles precipitation. Among precipitation method problems, one can name its required long time for product aging and washing process.

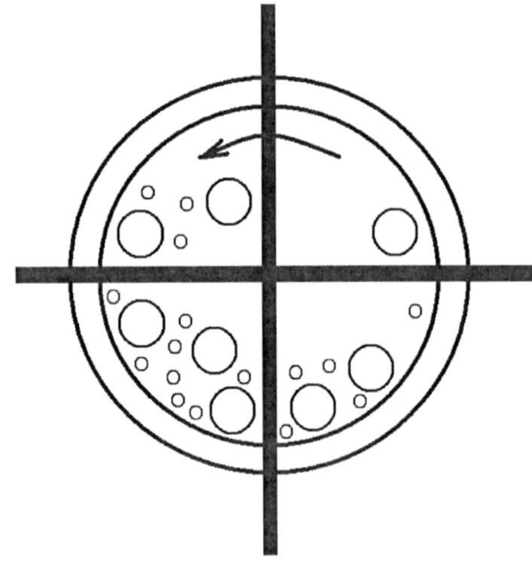

Figure 2.6. Milling process.

2.5.4. Sol-Gel Method

Sol-gel method is another chemical method for producing nano-materials. Sol-gel is a process for producing materials by developing sol, converting sol to gel, and finally solution exit. The solution can be each of Alco-oxide or salt combinations. Sol-gel is a chemical, or semi-chemical, method which is able to produce the desired product in a semi-stable (or amorphous) state, through creating a network by the relevant reactions. The point which gives priority for this method is the uniformity and homogeneity of the product - as one can say materials are combined with each other in molecular level.

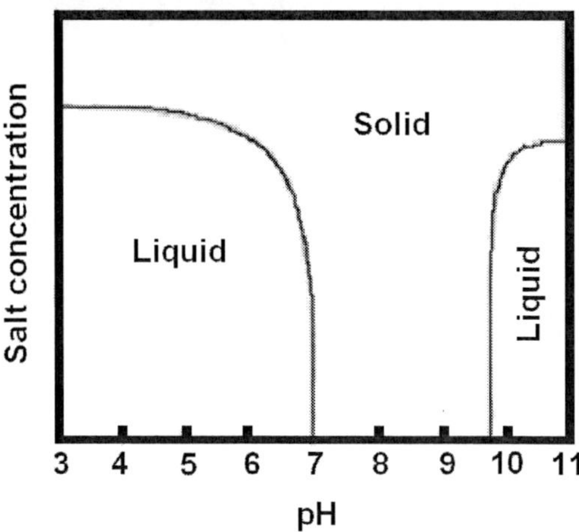

Figure 2.7. Alumina stability diagram.

2.5.4.1. Sol-Gel Concepts

Basically, a stable colloidal suspension composed of solid or polymeric dispersed particles in a liquid is named sol, where the particles can be amorphous or crystalline. The concept is different from that of Aerosol, which is particles suspension in a gas phase. A colloid is suspension, where the dispersed phase is as small (1 to 1000 nm) that gravity forces are ignored for them. The particles interactions are due to short-term forces, such as Van Der Waals gravity and surface charges. In these suspensions dispersed phase inertia is small enough, which leads to particles Brownic movements.

According a definition, gel is a persistent porous 3D network, which is accommodated in a liquid phase (wet gel). In colloidal or particulate gels, the network is developed through colloidal particles agglomeration. In polymeric gels, particles have a polymeric nano structure, which is made of small colloidal chemical unites agglomeration. In colloidal gels, sol particles are bounded via covalent boundaries, Van Der Waals forces, or hydrogen boundaries and construct gel structure; while polymeric gels make polymeric strings by entanglement.

Generally, sol-gel process is carried out in several stages, as follows:

1. Hydrolysis, molecular precursors condensation, and sol development
2. Sol conversion to gelatin

3. Aging
4. Drying

According these steps there would be a variety of products. As it can be understood, through this method one can create a coating with high condensation or control drying conditions and make a panel by applying an efficient thermal operation. Catalysts and powders can be produced on this basis, as we can create uniform distributed porosities in nano scales, which is of a great importance in producing catalysts and their bases. Although, controlling reactions conditions, there will be a potential for producing powders in different sizes and shapes, which implies the method's flexibility. Fibers are another product of this method, as one can extract a wanted fiber from developed sol and then apply a thermal operation on it.

One of important features of this method is the decrease in sintering temperature, caused by products high special surface. Alkoxides and salts are common rough materials for sol-gel process. Alkoxides can easily react with water and complete the required reactions for gelatinization process. Metallic alkoxides are among metal-organic combinations, which include an organic ligand bounded to a metal or pseudo-metal. The first metallic alkoxide was made by Elbemen through mixing alcohol and $SiCl_4$. He found that one can obtain the gelatin exposing the mixture to atmosphere. However these materials were of great interest for chemist for so many years till Geffeknen, in 1930, found that alkoxidecan be used in producing films and oxide thin layers.

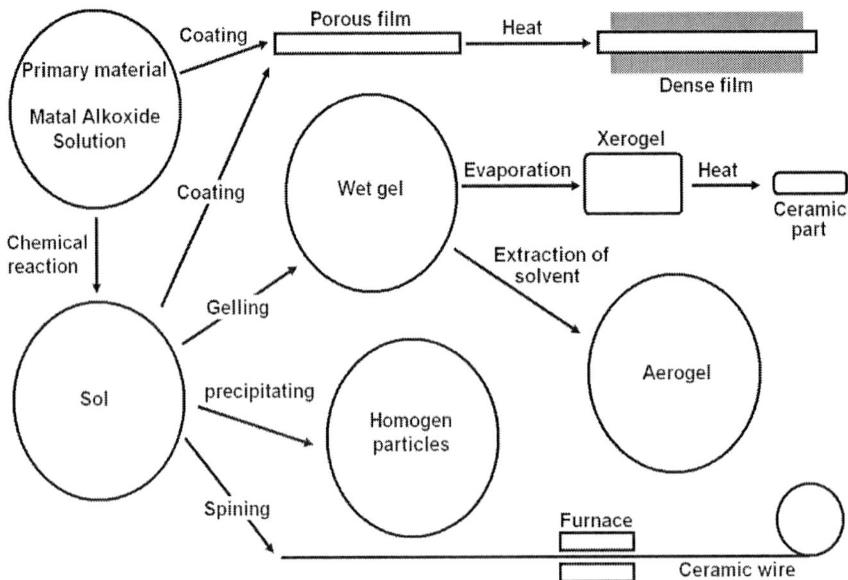

Figure 2.8. A variety of sol-gel products.

The process was developed by Schott Glass Company. However alkoxides high costs and being poisonous makes their uses be limited.

Salts are another group of rough materials for this method, which show less tendency for reaction compared with alkoxides and have a more difficult reaction control, however their costs are lower in comparison with alkoxides. Different steps for sol-gel process are introduced in following part.

2.5.4.2. Sol Development Physics

Sol's stabilization or coalescence is important in sol-gel process, while the obtained gel network and structure is related to agglomerated particles size and shape. Accurate controlling of intra-particle forces, we can achieve a colloidal suspension in a stable, weak follicular, and strong follicular state. Colloidal suspension stabilized state is where their potential energy increases through particles approaching to each other, and consequently detract each other in the suspension. In weak follicular state energy curve involves a minimum point. In this state agglomerated particles are in a balanced distance from each other, which in fact is the curves minimum distance.

When the potential energy decreases due to particles approaching to each other, there would be strong flocculation, which may lead to development of persistence network or separated clusters - which is in turn depended on concentration.

Colloidal stability is controlled by final intra-particle potential energy. The energy can be expressed as following:

$$V_{total} = V_{vdw} + V_{elect} + V_{steric} + V_{structural}$$

where:

V_{vdw} = Withdrawal potential energy caused by Van Der Waals interactions

V_{elect} = Detraction potential energy caused by electrostatic interactions between particles with similar charges

V_{steric} = Detraction potential energy caused by sterical interactions between coated particles surface in polymeric parts

$V_{structural}$ = Potential energy caused by particles presence which leads to a decrease or increase in suspensions stability

2.5.4.3. Salt Solutions Chemistry

For a salt system gelatinization process steps include hydrolysis and condensation reactions, which will be discussed as follows.

A. Hydrolysis

First, metallic cation composes a complex with water, due to salt dissolving in it.

In this state electron couple transmits from water molecule to empty orbitals of metallic cation. This leads to a slight increase in hydrogen charge and water molecule became fairly acidic. Based on acidity rate we can expect following hydrolysis products:

$$[M(OH_2)]^{Z+} = [M-O]^{(Z-1)+} + H^{+} = [M=O]^{(Z-2)+} + 2H^{+}$$

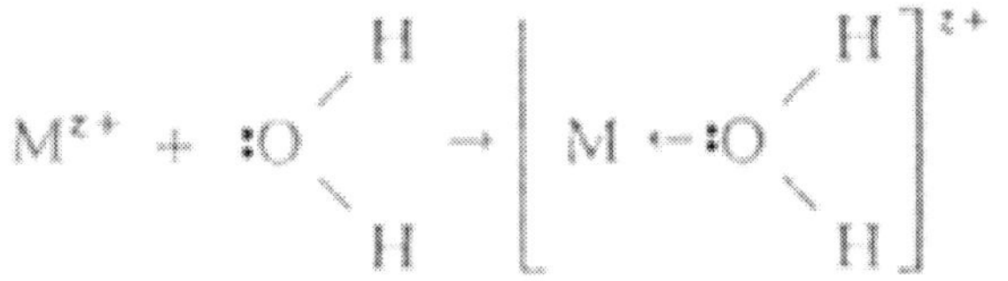

$$M^{z+} + :O\begin{matrix} H \\ H \end{matrix} \rightarrow \left[M \leftarrow :O\begin{matrix} H \\ H \end{matrix} \right]^{z+}$$

Figure 2.9. Development of metallic cation complex with water.

B. Condensation

Based on coordination number, condensation can be done through two different nucleophile mechanisms. Once coordination number of a given metal is known, condensation can be performed through a nucleophile substitution reaction:

$$M_1\text{-}OX + M_2\text{-}OY \longrightarrow M_1\text{-}\underset{\underset{X}{|}}{O}\text{-}M_2 + OY$$

When favorable coordination is not provided for the metal, it intends to obtain the required coordination number. As a result, condensation process is performed by an additive nucleophile reaction:

$$M_1\text{-}OX + M_2\text{-}OY \longrightarrow M_1\text{-}\underset{\underset{X}{|}}{O}\text{-}M_2\text{-}OY$$

The reaction above tends to increase coordination number of metal M2. Condensation reactions are divided into two classes, according their bound type:

I. Olation

Olation is a condensation process, through which hydroxide bridges are created between two metals. This reaction, as shown in figure 2.10, produce water and hydroxide group:

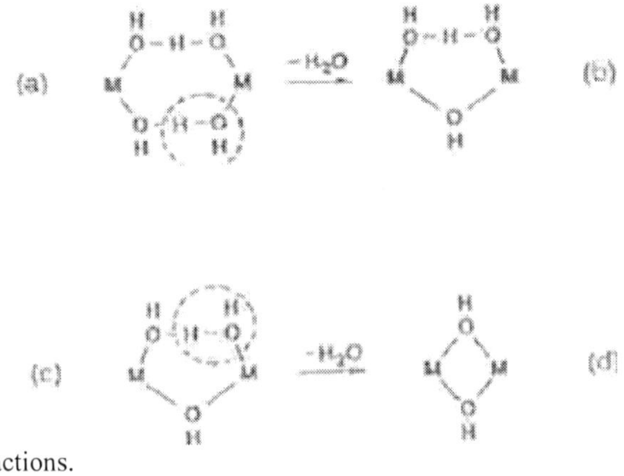

Figure 2.10. Olation reactions.

II. Oxolation

Oxolation is a condensation reaction, through which oxygen bridges are composed between two metals. Once coordination number of the metal is not saturated, the reaction is performed by nucleophile addition mechanism. It must be considered that, depending on the conditions, each oxolation and olation reactions can be occurred; even it is possible they happen simultaneously. As the condensation process is done, we will have hydroxide and oxygen bridges.

While metal coordination number is saturated, oxolation process is performed in two steps (figure 2.12), which produces water molecules.

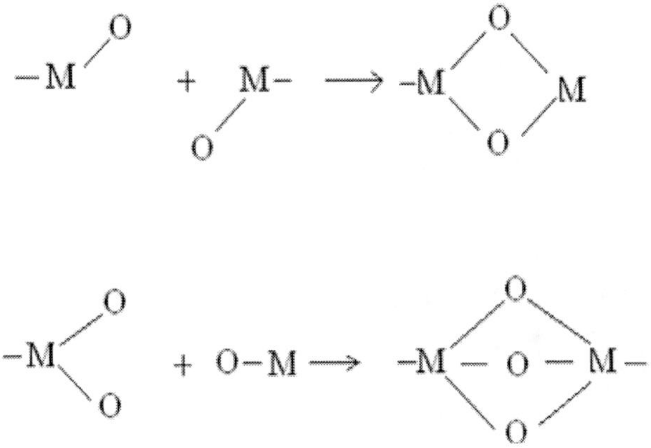

$$-M{-}O \;+\; O{-}M- \;\longrightarrow\; -M\langle{}^{O}_{O}\rangle M$$

$$-M\langle{}^{O}_{O} \;+\; O{-}M \;\longrightarrow\; -M\langle{}^{O}_{O}\rangle{-}O{-}M-$$

Figure 2.11. Oxolation reactions where metal coordination number is unsaturated.

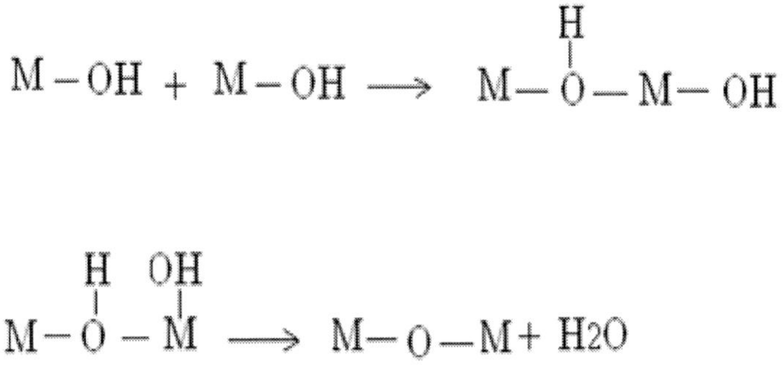

$$M-OH + M-OH \longrightarrow M-\overset{H}{\overset{|}{O}}-M-OH$$

$$M-\overset{H}{\overset{|}{O}}-\overset{OH}{\overset{|}{M}} \longrightarrow M-O-M + H_2O$$

Figure 2.12. Oxolation reaction where metal coordination number is saturated.

2.5.4.4. Chemistry of Metallic Alcoxide Solutions

According their different nature form that of salts, metallic alcoxides have different chemical reactions compared with them. These materials are intensely active, due to their strongly electronegative OR groups. M is in its highest oxidation state in these groups is very liable to nucleophile attacks. Gelatinization processes - hydrolysis and condensation basis is similar to that of salts.

When catalyst is absent and while metal coordination number is saturated, hydrolysis and condensation reactions are both performed through nucleophile substitution mechanisms, as adding a nucleophile group a proton transmits from invader molecule to alkoxide molecule and this protonech part separates in alcohol or water form. A schematic view of these interactions is shown in figure 2.13. Condensation, also, can be happened through olation reactions, illustrated in figure 2.14. After completing hydrolysis and condensation reactions gelatin, which creates a network and confines sol liquid, will be produced.

If gelatinization process of particulate sols is accompanied with creation of covalent inter-particle bounds, it would be irreversible, vice versa. In other words adding water and mixing the gelatin, bounds will be decomposed and sol will be developed. Basically, gelatinization length is decreased through all parameters which lead to an increase in speed of hydrolysis and condensation reactions.

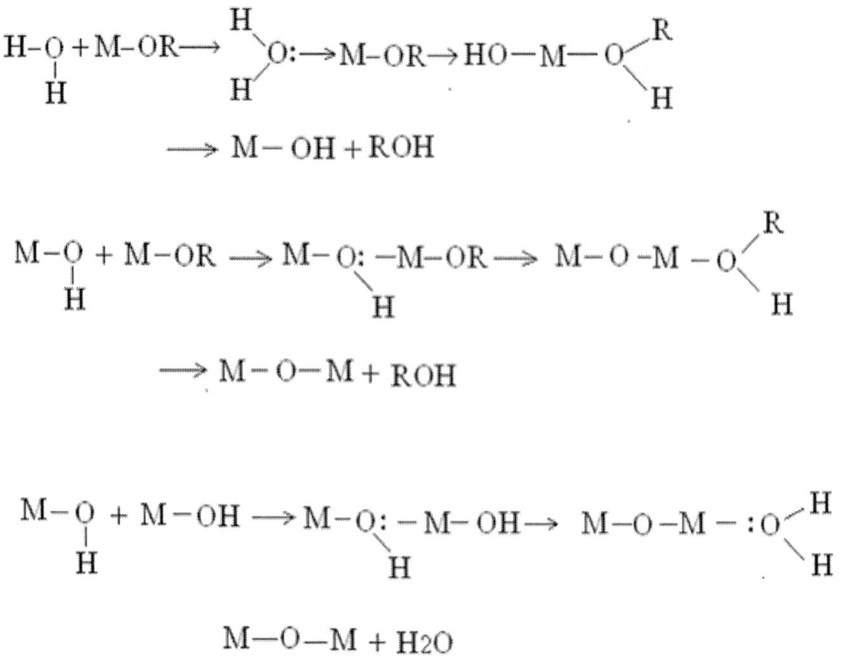

$$H{-}\underset{\substack{|\\H}}{O} + M{-}OR \longrightarrow \begin{matrix}H\\ \ \ \diagdown\\ \ \ \diagup\\ H\end{matrix}\!\!O{:}\!\longrightarrow M{-}OR \rightarrow HO{-}M{-}O\begin{matrix}\diagup R\\ \diagdown H\end{matrix}$$

$$\longrightarrow M-OH+ROH$$

$$M{-}\underset{\substack{|\\H}}{O} + M{-}OR \longrightarrow M{-}\underset{\substack{\diagdown\\H}}{O}{:}\,{-}M{-}OR \longrightarrow M{-}O{-}M{-}O\begin{matrix}\diagup R\\ \diagdown H\end{matrix}$$

$$\longrightarrow M-O-M+ROH$$

$$M{-}\underset{\substack{|\\H}}{O} + M{-}OH \longrightarrow M{-}\underset{\substack{\diagdown\\H}}{O}{:}\,{-}M{-}OH \longrightarrow M{-}O{-}M{-}{:}O\begin{matrix}\diagup H\\ \diagdown H\end{matrix}$$

$$M-O-M+H_2O$$

Figure 2.13. Hydrolysis and condensation mechanisms (oxolation).

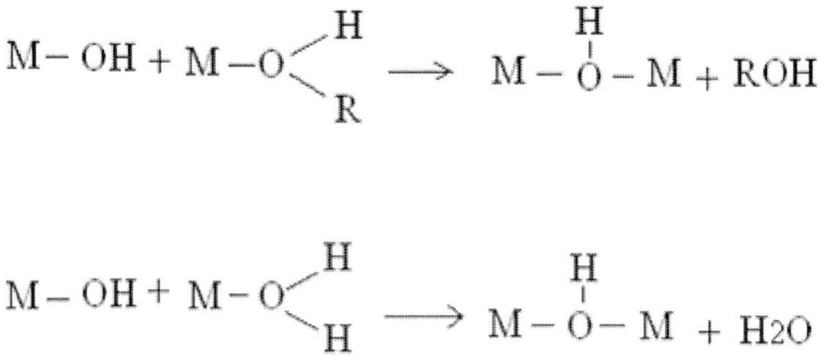

$$M-OH+M-O\begin{matrix}\diagup H\\ \diagdown R\end{matrix} \longrightarrow M-\overset{\substack{H\\|}}{O}-M+ROH$$

$$M-OH+M-O\begin{matrix}\diagup H\\ \diagdown H\end{matrix} \longrightarrow M-\overset{\substack{H\\|}}{O}-M+H_2O$$

Figure 2.14. Olation reactions.

2.5.4.5. Aging

A strong increase in viscosity of gelatinization point makes us think gelatins network structure is stabilized. In spite of this structure, it is probable that pores liquid considerably changes, depending on temperature, pH, solution conditions. Chemical reactions are not completed with gelatinization, so structural rearrangement in wet gelatins (one with pore space liquid) will occur. This process causes an increase in gelatins hardness, known as aging.

2.5.4.6. Drying

Achieving a uniform gelatin one must carefully perform drying operation. If drying is simply done through evaporation of the solution from the body, heating speed has to be as low as possible to prevent stress development and, consequently, body damage. Drying via pore space liquid evaporation leads to contraction of gelatin network by creating a viscose force. Regarding pores' different sizes and, in turn, results of different forces application, one can expect crack developments on walls (figure 2.15).

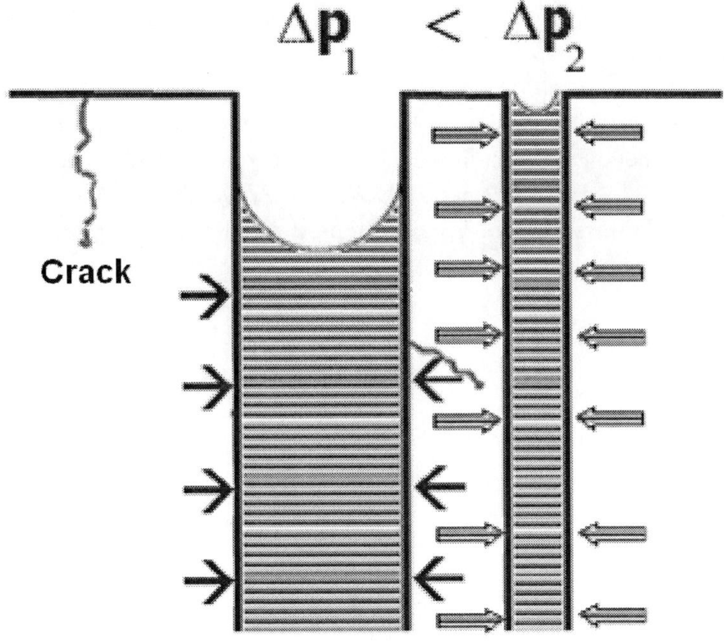

Figure 2.15. Crack development during drying process.

A gelatin obtained by drying process is known as xerogel. It typically has lower volume compared with primary gelatin. A common method for drying gelatinized walls is creating saturation around the walls and steady exit of vapor. An efficient method for gelatins drying is solution exit under hyper-critical conditions. Gelation obtained in this method is submerged in a solvent, such as ethyl acetate, and this set is put in autoclave. Then in sample content, carbon dioxide is added in room temperature until it is substituted with solvent. Room temperature is gradually increased to slowly let the carbon dioxide out. The gelatin obtained in this method is known as aerogel. As there is no evidence of contraction in this gelatin, it is very porous and has a fairly low density, but there are no stresses caused by contraction. Applying thermal operations on aerogels or xerogels one can produce glasses and ceramics with high density. Among other hypercritical drying methods we can name freezing drying. Through this method, freezing the intra-sample liquid and creating the vacuum we can evaporate the liquid.

References

[1] Adeloju, S.B. and Ohanessian, A. Layer by layer fabrication of an amperometric nanocomposite biosensor for sulfite. 2006 NSTI Nanotechnology Conference and Trade Show - NSTI Nanotech 2006 Technical Proceedings, pp. 190-1912006).

[2] Aliofkhazraei, M. and Sabour Rouhaghdam, A. Fabrication of TiC/WC ultra hard nanocomposite layers by plasma electrolysis and study of its characteristics. *Surface and Coatings Technology*, 2010.

[3] Bo, Y., Lei, S., Nangeng, W., Xiaohan, L., Limin, W. and Jian, Z. A facile method for fabrication of ordered porous polymer films. *Macromolecules*, 2008, 41(18), 6624-6626.

[4] Cao, H., He, J., Deng, L. and Gao, X. Fabrication of cyclodextrin-functionalized superparamagnetic Fe3O4/amino-silane core-shell nanoparticles via layer-by-layer method. *Applied Surface Science*, 2009, 255(18), 7974-7980.

[5] Chraska, T., Neufuss, K., Dubsky, J., Ctibor, P. and Klementova, M. Fabrication of bulk nanocrystalline ceramic materials. *Journal of Thermal Spray Technology,* 2008, 17(5-6), 872-877.

[6] Chraska, T., Neufuss, K., Dubsk¡½, J., Ctibor, P. and Rohan, P. Fabrication of bulk nanocrystalline alumina-zirconia materials. *Ceramics International*, 2008, 34(5), 1229-1236.

[7] Das, R.N., Lauffer, J.M. and Markovich, V.R. Fabrication, integration and reliability of nanocomposite based embedded capacitors in microelectronics packaging. *Journal of Materials Chemistry*, 2008, 18(5), 537-544.

[8] Fan, Y., Wang, R. and Moradian-Oldak, J. Fabrication of enamel-mimicking mineralization composite coating induced by electrolytic deposition (ELD) system. *Materials Research Society Symposium Proceedings*, pp. 12-172007).

[9] Fan, Y.W., Sun, Z., Wang, R., Abbott, C. and Moradian-Oldak, J. Enamel inspired nanocomposite fabrication through amelogenin supramolecular assembly. *Biomaterials,* 2007, 28(19), 3034-3042.

[10] Garcia, E.J., Wardle, B.L., John Hart, A. and Yamamoto, N. Fabrication and multifunctional properties of a hybrid laminate with aligned carbon nanotubes grown In Situ. *Composites Science and Technology*, 2008, 68(9), 2034-2041.

[11] Gutierrez-Tauste, D., Domenech, X., Casan*f*-Pastor, N. and Ayllon, J.A. Characterization of methylene blue/TiO_2 hybrid thin films prepared by the liquid phase deposition (LPD) method: Application for fabrication of light-activated colorimetric oxygen indicators. *Journal of Photochemistry and Photobiology A: Chemistry*, 2007, 187(1), 45-52.

[12] Ha, Y.G., You, E.A., Kim, B.J. and Choi, J.H. Fabrication and characterization of OLEDs using MEH-PPV and SWCNT nanocomposites. *Synthetic Metals*, 2005, 153(1-3), 205-208.

[13] He, L.P., Mai, Y.W. and Chen, Z.Z. Fabrication and characterization of nanometer CaP(aggregate)/Al_2O_3 composite coating on titanium. *Materials Science and Engineering* A, 2004, 367(1-2), 51-56.

[14] Huang, C.C. The mechanical fabrication of nanocomposites and engineered particles. 2007 NSTI Nanotechnology Conference and Trade Show - NSTI Nanotech 2007, *Technical Proceedings*, pp. 677-6792007).

[15] Ii, Zhang, R., Wang, H., Gao, L., Guan, S. and Guo, J. Spheroid growth during sintering of copper coated silicon carbide particles in the fabrication of nanocomposite. *Key Engineering Materials*, pp. 1275-12782005).

[16] Jiang, L. and Gao, L. Fabrication and characterization of ZnO-coated multi-walled carbon nanotubes with enhanced photocatalytic activity. *Materials Chemistry and Physics*, 2005, 91(2-3), 313-316.

[17] Jiang, P. and McFarland, M.J. Large-scale fabrication of wafer-size colloidal crystals, macroporous polymers and nanocomposites by spin-coating. *Journal of the American Chemical Society*, 2004, 126(42), 13778-13786.

[18] Kim, S.H., Kim, M.J. and Choa, Y.H. Fabrication and estimation of Au-coated Fe_3O_4 nanocomposite powders for the separation and purification of biomolecules. *Materials Science and Engineering* A, 2007, 448-451, 386-388.

[19] Kong, E.S.W., Fang, F., Zhang, Y., Qian, S., Tang, B.Z., Wang, Z.L. and Wang, X. Nonlinear optical polymers and nanophtotonic devices: Nanofabrication of devices based on nanocrystals-embedded polymers, UV-assisted nanoimprinting, and ring-resonator planar lightwave circuitry. *Proceedings of SPIE - The International Society for Optical Engineering* 2005).

[20] Krzanowski, J.E. and Zimmerman, J.H. Fabrication and tribological properties of hard coatings with embedded solid lubricant phases. *Proceedings of STLE/ASME International Joint Tribology Conference,* IJTC 20062006).

[21] Lee, S.H., Ham, H.T., Park, J.S., Chung, I.J. and Kim, S.O. Fabrication of ordered porous SWNT-polymer nanocomposites by emulsion templating. *Macromolecular Symposia*, 2007, 249-250, 618-622.

[22] Li, B., Li, Y. and Zhang, J. Fabrication of hard/soft nanocomposite magnetic particles by sonochemistry. Xiyou Jinshu Cailiao Yu Gongcheng/*Rare Metal Materials and Engineering*, 2007, 36(8), 1457-1460.

[23] Li, Y., Zhang, J., Qiao, G. and Jin, Z. Fabrication and properties of machinable 3Y-ZrO_2/BN nanocomposites. *Materials Science and Engineering* A, 2005, 397(1-2), 35-40.

[24] Liu, A., Li, D., Honma, I. and Zhou, H. Multilayered mesoporous titanate nanocomposite film: Fabrication by layer-by-layer self-assembly and its electrochemical properties with H + intercalation. *Electrochemistry Communications*, 2006, 8(2), 206-210.

[25] Liu, X., Wang, J., Zhang, J., Liu, B., Zhou, J. and Yang, S. Fabrication and characterization of Ag/polymer nanocomposite films through layer-by-layer self-assembly technique. *Thin Solid Films*, 2007, 515(20-21), 7870-7875.

[26] Lu, H., Liu, Y., Gou, J. and Leng, J. Fabrication and properties of shape-memory polymer coated with conductive nanofiber paper. Proceedings of SPIE - *The International Society for Optical Engineering*2009).

[27] Pang, X. and Zhitomirsky, I. Fabrication of chitosan-hydroxyapatite coatings for biomedical applications. *ECS Transactions*, pp. 15-222007).

[28] Perlich, J., Memesa, M., Diethert, A., Metwalli, E., Wang, W., Roth, S.V., Timmann, A., Gutmann, J.S. and Muller-Buschbaum, P. Layer-by-layer fabrication of hierarchical structures in sol-gel templated thin titania films. *Physica Status Solidi - Rapid Research Letters*, 2009, 3(4), 118-120.

[29] Psuja, P., Hreniaka, D., Streka, W., Kovalenkob, D. and Gaishunb, V. Fabrication and properties of nanocomposite ITO layer containing terbium doped yttrium aluminum garnet nanocrystallites. *Proceedings of SPIE - The International Society for Optical Engineering*2009).

[30] Qi, Z.M., Honma, I., Ichihara, M. and Zhou, H. Layer-by-layer fabrication and characterization of gold-nanoparticle/ myoglobin nanocomposite films. *Advanced Functional Materials,* 2006, 16(3), 377-386.

[31] Shaikh Mohammed, J., DeCoster, M.A. and McShane, M.J. Fabrication of interdigitated micropatterns of self-assembled polymer nanofilms containing cell-adhesive materials. *Langmuir*, 2006, 22(6), 2738-2746.

[32] Sun, P., Jiang, Y., Xie, G., Yu, J., Yu, Y. and Hu, J. Fabrication and gas sensitivity of poly-2, 5-dimethoxyethynylbenzene/ SnO2 nanocomposite. *Proceedings of SPIE - The International Society for Optical Engineering*2009).

[33] Tang, E., Liu, H., Sun, L., Zheng, E. and Cheng, G. Fabrication of zinc oxide/poly(styrene) grafted nanocomposite latex and its dispersion. *European Polymer Journal*, 2007, 43(10), 4210-4218.

[34] Tian, Y.S., Chen, C.Z., Wang, D.Y., Chen, L.X. and Lei, T.Q. Research progress in fabrication of ceramic coatings. Cailiao Kexue yu Gongyi/*Material Science and Technology*, 2005, 13(4), 427-430+434.

[35] Venkatesh, S., Jiang, P. and Jiang, B. Generalized fabrication of two-dimensional non-close-packed colloidal crystals. *Langmuir*, 2007, 23(15), 8231-8235.

[36] Vickery, J.L., Patil, A.J. and Mann, S. Fabrication of graphene-polymer nanocomposites with higher-order three-dimensional architectures. *Advanced Materials*, 2009, 21(21), 2180-2184.

[37] Wang, X., Qiao, G. and Jin, Z. Fabrication of machinable silicon carbide-boron nitride ceramic nanocomposites. *Journal of the American Ceramic Society,* 2004, 87(4), 565-570.

[38] Wang, Z., Chen, X., Chen, M. and Wu, L. Facile fabrication method and characterization of hollow Ag/SiO2 double-shelled spheres. *Langmuir,* 2009, 25(13), 7646-7651.

[39] Wu, Z., Wu, D., Li, F., Hong, F., Qi, S. and Jin, R. Surface modification-based fabrication of double surface highly reflective and conductive metallized polymeric films and microstructural tuning nanocomposite layers. *Advanced Materials Research*, pp. 497-5002006).

[40] Xu, W., Chen, H., Li, H. and Wang, M. Fabrication of carbon black/crosslinked poly(vinyl pyrrolidone) core-shell nanoparticles stable in water. *Colloids and Surfaces A: Physicochemical and Engineering Aspects*, 2005, 266(1-3), 68-72.

[41] Xu, X., Zhang, Z. and Liu, W. Fabrication of superhydrophobic surfaces with perfluorooctanoic acid modified TiO_2/polystyrene nanocomposites coating. *Colloids and Surfaces A: Physicochemical and Engineering Aspects,* 2009, 341(1-3), 21-26.

[42] Yao, K.X. and Zeng, H.C. Fabrication and surface properties of composite films of SAM/Pt/ZnO/SiO_2. *Langmuir*, 2008, 24(24), 14234-14244.

[43] Zeng, Z., Zhou, Y., Zhang, B., Sun, Y. and Zhang, J. Designed fabrication of hard Cr{single bond}Cr_2O_3{single bond}Cr_7C_3 nanocomposite coatings for anti-wear application. *Acta Materialia*, 2009, 57(18), 5342-5347.

Chapter 3

Basic Characterization of Nanocomposite Coatings

Abstract

After introducing nanocomposite coatings and their fabrication methods in previous chapters, this chapter discusses about basic characterization of nanocomposite coatings. It starts by electrodeposition method and continues by some examples about polymeric-based nanocomposite coatings. Effects of some important factors on coating method have been discussed. Some classifications about nanocomposite coatings were also mentioned. Although the starting point of composite coatings production was in early 19th century, most applications and models of producing composite coatings have been proposed during past forty years. Coatings with nickel matrix which contain silicon carbide particles were used in 1966 with the aim of industrial application in producing rotors, which was the first industrial application of composite coatings. Because of successful electroplating of composite coatings, similar technologies were developed for creating electroless composite coatings. Advanced physical properties of such coatings helped employ them for newer applications such as using the coatings in electro-catalysts and photo-electro-catalysts. By the advent of nano-materials during last decade, electro-deposition methods, set the stage for obtaining different products, e.g. nanostructured coatings, nano-wires, nano-tubes, nanometer multi-layers and nano-composites, from nanostructured materials. Most recent research have focused on strengthened composite coatings, high electrical resistance in printed circuit boards, appropriate magnetic resistance in data-saving systems, and rigidity of micrometric devices used in microelectronic systems (MEMS).

3.1. Electroplating of Composite Coatings

Electroplating of composite coatings involves the processes in which micron-scale or even smaller particles are suspended in an electrolyte. Such particles are placed in a solid context generated by an electrical current, and they confer special characteristics to the matrix, depending on quantity, type, and distribution of the particles. Electro-deposition has some advantages over other competing methods; the advantages of electroplating are as follows:

- High production rate
- Low limitation of size and shape
- Availability of industrial equipments
- Fewer obstacles in transferring technology from the research laboratory to industry
- Capability of producing pure metals as well as basic alloys and metal composites through one run of electroplating process
- Competency for creating fully-compressed, non-porous nanostructures

In present trend toward producing miniature pieces, electroplating has been proposed as a potential technology. Using electroplating, metals and composites can generated for microelectronic systems and systems with high volume-to-surface ratio. These microsystems have a height of about some hundred micrometers and a thickness of some micrometers to tens of micrometers, and can be made of materials such as metals, ceramics, and polymers. Having large surfaces (between tens of square centimeters and some square meters), these materials are applicable in different fields, e.g. heat transfer, fluids mechanics, catalytic systems, as well as producing composite materials headstocks. Using nanostructures in creating composites of metal context is of main interest in microelectronic systems (MEMS). Being highly rigid and resistant to wearing and corrosion makes Nano-composite materials a suitable option in producing next-generation microsystems. For manufacturing micrometer devices, the composite material must possess particles at the scale of nanometer.

Table 3.1. Different types of nano-structured materials which can be produced through electroplating

<table>
<tr><td rowspan="2">Methods of electrodeposition</td><td colspan="4">Type of nanostructured materials</td></tr>
<tr><td>Nanoparticles in a metallic matrix</td><td>Nano-multilayer</td><td>Nanotubes/ nanowires</td><td>Nanocrystalline materials</td></tr>
<tr><td>Direct current (DC)</td><td>Ni/Alumina</td><td rowspan="5">Ni-Cr</td><td rowspan="5">Co nanowires</td><td rowspan="5">Ni-W alloy</td></tr>
<tr><td>Pulsed direct current (PDC)</td><td>Ni-W/CNT</td></tr>
<tr><td>Pulsed reverse current (PRC)</td><td rowspan="3">Multilayered composites</td></tr>
<tr><td>Potentiostatic (P)</td></tr>
<tr><td>Pulsed potetiostatic (PP)</td></tr>
</table>

3.2. The History of Composite Coatings Produced by Electroplating

Although the starting point of composite coatings production was in early 19th century, most applications and models of producing composite coatings have been proposed during past forty years. In 1928, copper-graphite codeposition was examined for generating self-lubricating coatings of car engines. In addition, a patent was filed in 1962 on deposition of different types of particles in metallic matrix. Coatings with nickel matrix which contain silicon carbide particles were used in 1966 with the aim of industrial application in producing rotors, which was the first industrial application of composite coatings. In 1970s, usage of composite coatings such as Co-WC helped enhance the resistance against wear and stiffness in high temperatures. Because of successful electroplating of composite coatings, similar technologies were developed for creating electroless composite coatings. Advanced physical properties of such coatings helped employ them for newer applications such as using the coatings in electro-catalysts and photo-electro-catalysts. By the advent of nano-materials during last decade, electro-deposition methods, set the stage for obtaining different products, e.g. nanostructured coatings, nano-wires, nano-tubes, nanometer multi-layers and nano-composites, from nanostructured materials. Most recent research have focused on strengthened composite coatings, high electrical resistance in printed circuit boards, appropriate magnetic resistance in data-saving systems, and rigidity of micrometric devices used in microelectronic systems (MEMS). Table 3.1 shows different groups of nano-structured materials and various methods for their production.

3.3. THE MECHANISM OF ELECTROCHEMICAL CODEPOSITION

The mechanism of particle codeposition remains to be fully understood. Adsorption and electrophoretics have been considered as the key parameters in the proposed mechanisms. Since weight percentage of the particles in a coating depends on the amount of metal deposition, both deposition rates of particles and the metal should be considered.

Electrophoresis is passage of charged particles through a solution by employing an electrical field. In 1962, the researchers attempted to elucidate the reason of codeposition. They claimed that displacement of charged particles toward cathode by electrophoresis is the reason for particles codeposition. The important fact is that ionic strength is high in most electroplating solutions, thus the displacement caused by electrophoresis is limited due to the small potential gradient in the solution.

In 1964, it was put forward that particle displacement happens as a result of solution turbulence, so that the number of particles colliding with anode surface increases as the solution gets more turbulent. In addition, it was claimed that particles codeposition is controlled by particles contact time with cathode and the rate of metal deposition.

A model was developed in 1967 in which particles are placed in the coating under the effect of Van Der Waals force interaction between the particles and cathode surface. Therefore, the particles are buried in growing metallic matrix when they are adsorbed into the cathode. Another model was provided in 1968 which was capable of determining the weight percentage of the particles present in the coating. However, this model was discredited due to being solely based on mechanical entrapment of the particles. In 1971, a group of researchers claimed that considering the turbulence, the particles must stay on cathode surface for a definite time to be able to enter the coating. They provided an equation for determining particles weight percentage in the coating, although it was doubtful because its estimated particle weight percentages were not consistent with the real percentages.

Guglielmi in 1972 proposed a mechanism which was a big step toward clarifying particles codeposition mechanism. This model has been employed for justifying the mechanism of composite electroplating of Ni/TiO_2 and Ni/SiC. The sizes of the particles used were between 1 to 2 micrometers and current density between 2 and 10 ampere per square decimeter. This model is generally constituted of two consecutive stages. In the first stage, a weak physical adsorption of the particles on cathode surface occurs. The weakly adsorbed particles are surrounded by ions. In the second stage, which is an electrochemical stage, the particles are adsorbed to the cathode by the cathode electrical filed and are buried by reducing their surrounding ions in the growing metallic layer.

A mathematical formula for determining the rate of particle entrapment was developed by researchers in 1974 and based on two experimental observations. The first evidence indicated that increasing the rate of turbulence causes a raise in particle collision with cathode surface and bases the second evidence, the mechanical force exerted on the particles placed in cathode surface is affected by turbulence.

Celis, Roos, and Buelens in 1987 attempted to predict the amount of codeposited particles in metal-particle system by formulating a model which analyzed particles transfer from solution volume to the cathode surface. They employed this model in explaining the mechanism of composite electroplating of Cu/Al_2O_3 and Au/Al_2O_3. In their analysis, the size of the particles was 50 nm, current density 0 to 9 A/dm^2, and turbulence speed 400 to 600 rounds per minute. The concept of possibility was used in this model to show the quantity of the particles that can be deposited by a definite current density. They provided following five-staged process for particles codeposition in metals: in the first stage of their proposed process, the particles present in the solution adsorb ions from the solution on their surface, so that a double layer is created around each particle; in second step, the particle reach the marginal layer through displacement; the particles move from the diffusion layer toward the cathode. Finally, the particles enter the coating in two steps which are similar to the model developed by Guglielmi. The particles surrounded by ions are adsorbed into the cathode, and then they become buried in the coating while some of the ions on their surface are reduced.

Some other researchers provided a more general model than that of Celis. This model is based on Guglielmi's model and states that the rate of particles codeposition is determined by the reduction rate of the ions adsorbed on the particles, which depends on the diffusion and kinetic parameters. Codeposition of Co-SiC has been examined in this model, and the model focused on reduction of hydrogen and cobalt ions. In reduction of the ions adsorbed on the cathode, three ranges of current density have been selected: low current density in which only hydrogen ion can be reduced, medium current density in which cobalt ion is reduced and hydrogen ion reaches a limiting current density, and high current density in which reduction of both the ions is under the control of their displacement.

Another group of researchers developed a model in which mass displacement was corresponding to a series of numbers. Such numbers were without dimension and were determined by the factors affecting mass displacement. Sherwood number, used for particle displacement toward a stable plate in a dilute solution, was optimized for particles codeposition. However, this model is not capable of predicting key parameters such as particles concentration in the coating according to the current density since a limited range of experimental conditions were assumed for this model and the influence of kinetic factors were ignored.

In 1987, a model was proposed for particle codeposition at nanometer scale on Rotating Disk Electrode (RDE), in which different ways of particles displacement has been studied. A continuity equation was described for particle concentrations in proportion to mass difference balance. This term included different processes of mass transfer, such as external diffusion and displacement. It was suggested that full immersion model predicted maximum particle concentration in a limiting current density that was against the experimental observations. Full immersion model indicates that all the particles placed at a definite distance from cathode enter the coating. Therefore, there was a demand for a model which is corresponding to the experimental observations, and in which deposition rate of ions adsorbed on the particles is a determinative factor in particles codeposition.

A different model for explaining copper based composite codeposition system was proposed. In this model particles size was 11μm, current density 0 to $8A/dm^2$ and turbulence speed up to 700 rpm. In addition, in Ni/SiC system, the particles size was 10 nm to 10μm, current density 0 to $20A/dm^2$, and agitation rate 0 to 2000 rpm. They provided Trajectory model for determining particle trajectories. This model was based on all the forces and torques acting on the particles, including gravity, electrophonic forces, and the forces originated from the double layer. This model was not consistent in proximity of the electrode since it caused a full immersion. To solve this problem, they included in their equations a term representing the interactions between particles and electrodes. Their new equation for possibility of particles entry in the coating provided a better explanation for the change in particle concentration in the coating based on particle concentration in the solution. Trajectory model is not able to predict the maximum concentration of the particles present in the coating in proportion to the changes in current density.

A newer model for Ni/Al_2O_3 system was developed in 2000; size of the particles used in this model was 300nm, current density 0.5 to 4 A/dm^2, and turbulence speed 500 to 2000 rpm. In this model, particle displacement toward the surface is controlled by diffusion-convection. The effects of particle weight and solution turbulence have been considered in different current densities. This model is only creditable when particle sizes are smaller than the thickness of the diffusion layer.

Another model for Ni/PTFE system was employed in 2002, in which particle size was 500nm. In comparison to Guglielmi's model, this was a more advanced one, in which a modifying factor (a third order equation) is used for evaluating the effects of adsorption and turbulence condition. The mechanism developed by Guglielmi has been used by many researchers for elucidating deposition behavior of composite coatings. This model is capable of evaluating the effects of particle concentration and current density on amount of particles in the coating, although the effects of turbulence and particle properties have been ignored. In the modifying factor (the third order equation) in Guglielmi's model was used for considering the effects of adsorption and turbulence on presence of 500 nm PTFE particles in nickel coating. However, this model is creditable within only special experimental conditions and needs to be thoroughly studied for verifying its capability of predicting other systems.

Although many theoretical models have been developed so far, efficiency of each of them for assessing behavior of wider range of metallic coatings and various particles needs to be studied. Most available models include mathematical relations which are present among a series of parameters. In none of the models proposed by earlier authors, the effect of the particles on grain electrical budding has been analyzed. Considering the effective parameters such as particle properties (type and size), working condition (temperature, current density and pH), and electrolyte composition (concentration, presence of surfactants and additives) is a basic demand in future models.

Guglielmi's model provides a mechanism for creating metal-particle composite coating using a direct current. We describe this model in this section to elucidate some problems in composite coating codeposition. Note worthily, the effect of turbulence has not been considered in this model, which is one of its disadvantages. This model pursues two general objectives:

1. Studying the effect of the particles present in bath on the amount of particle entry into the metallic coating
2. Determining the amount of the particles present in the coating as a function of current density

Curve of α versus C in which α is the volume fraction in the coating and C the particle concentration in the bath is experimentally and theoretically important. According to adsorption theory by Langmuir, α is increasing proportionally to C until it reaches a saturation limit. Based on Langmuir's adsorption theory and Faraday laws, Guglielmi developed following equation:

$$\frac{C}{\alpha}=\frac{Mi_0}{nF\rho V_0}e^{((A-B)\eta)}.(C+\frac{1}{k})$$

where, α is the volume fraction of the deposited particles, C is the volume percentage of the particles suspending in the electroplating bath, M is atomic weight of the deposited metal, i_0 is the exchange current density, n is the valence of the deposition metal, F is faraday constant, ρ is density of deposition metal, and η is the over-potential for electrode reaction. K is the Langmuir constant which is determined by the intensity of particle collision with the cathode. Constants V_0 and B are relevant to particle deposition and constant A to metal deposition.

For coatings created in variable amounts of current density, C/α curve provide direct lines all passing from a common point on C-axis (x-axis). According to the above equation, when C/α is zero, quantity of -1/K is obtainable, which gives the quantity of Langmuir's adsorption constant. In fact, K is in the form of $K = K_a/K_d$, where K_a is adsorption coefficient and K_d is the repulsion coefficient. When K>1 the rate of particle adsorption is higher than the rate of particle repulsion, and vice versa. The size of PTFE particles used is 0.5 μm, current density between 1 and 7 A/dm^2, the velocity of solution rotation (originated by magnet) 500 rpm, the amount of PTFE present in the bath 30 g/l (1 percent in volume). Through extrapolation of the lines drawn, the amount of K is 1.9. K has also been found for large silicon carbide particles (2 μm) to be 8.3. Having K, the concentration of the particles weakly adsorbed on cathode surface (σ_1) and the concentration of the particles firmly adsorbed on it (σ_s) is determined using Langmuir's formula, which has been optimized by Guglielmi for electrical codeposition systems.

Table 3.2. volume percentages for particles weakly or firmly adsorbed on cathode surface during generation of nickel-silicon carbide composite coating

C%	σ_1	σ_s
0.1	17.3	1.76
0.3	37.5	3.36
0.5	48.9	5.38
1	63.5	6.7
1.5	70.6	7.46

Assuming all the particles firmly adsorbed on the cathode are buried in the coating ($\sigma_s = \alpha$), the percentage of the weakly-absorbed particles is determined through following equation. Table 3.2 presents volume percents for both weakly- and firmly-adsorbed particles of nickel-silicon carbide composite coating, containing nano-metric silicon carbide particles (with dimension of 62 nm) generated by direct current.

$$\sigma_1 = \frac{kC}{1+kC}(1-\sigma_s)$$

As it can be seen from table 3.2, the concentration of the particles weakly attached to the cathode is much higher than particle concentration inside the solution. In addition, only 0.1 percent of the particles weakly-adsorbed on the cathode can enter into the coating. One might therefore conclude that the speed determination step for particles participation is controlled by the process in which a weak adsorption is converted to a firm one.

Using estimation, cathodic current density is obtainable through $i_c = i_0 e(A\eta)$ and the quantity of B/A is determined as follows:

$$i_0.e^{(A-B).\eta} = i_0.e^{(A-B/A).A.\eta} = i_0^{(B/A)}.i_c^{(1-B/A)}$$

Combing two previous equations, we get

$$\alpha = \frac{nF\rho V_0}{Mi_0^{B/A}}.\frac{C}{1/k+C}.i_0^{(B/A-1)}$$

By calculating logarithm of above equation, a linear relation between log α and log i_c is obtained:

$$\log\alpha = \log\left(\frac{nF\rho v_0}{Mi_0^{B/A}}\right) + \log\left(\frac{C}{1/k+C}\right) + (B/A-1)\log(i_c)$$

Employing experimental results, curve of log α against logi_c for Ni-PTFE system has been drawn. All the curves were parallel to each other and so B/A ratio must be independent from PTFE concentration in the solution. The amount obtained for B/A of this system is 0.66. Two sets of electrochemical reactions generally occur in composite electroplating. The first set is pertinent to free metal ions passed Helmholtz double layer and reduced and deposited on cathode surface. The second set presents the particles which adsorb ions form the solution, pass the double layer, reach the cathode, and are finally buried in the growing coating because of reduction of the ions present on their surface. Constant A and B represent total charges displaced by both free ions and the ions adsorbed on the particles. If A>B, the rate of particle presence in the coating will be reduced as the current density increases, and vise versa.

3.4. Colloidal Systems Stability

Generally, colloidal systems stability, which is controlled by total inter-particle energy, can be expressed mathematically as follows:

$$V_{total} = V_{vdw} + V_{elect} + V_{steric} + V_{structural}$$

where V_{vdw} is the attracting potential energy because of Van Der Waals interaction between the particles, V_{elect} is the repulsion potential energy created from electrostatic interactions between equally-charged surfaces of the particles, V_{steric} is the repulsion between particle surfaces that have been covered by additives, and $V_{structural}$ is the potential energy originated from presence of unadsorbed components on the particles inside the solution, which might increase or decrease the suspension stability.

Generally, the most important factors affecting particle distribution in the solution are surface charge of suspending particles and ionic solidity of the solution in which the particles are suspending. The quantity of the ratio of these two parameters determines how the particles interact with each other. When the ratio of surface charge of the particles and ionic solidity is high, the result of the interaction between the particles is repulsion, and consequently a low-viscosity solution. Conversely, when quantity of the ratio is low, the particles will be in agglomerated form inside the solution. The effect of ionic power of the solution on agglomeration of silicon carbide particles (250 nm) illustrated that agglomeration rate of silicon carbide particles in distilled water containing 0.1 g/l silicon carbide and without using additives, is much lower that in nickel electroplating bath (sulfamate bath) containing the same amount of silicon carbide under similar conditions. This observation is pertinent to the higher ionic power of the electroplating solution than that of water.

When there are many equally-charged particles, they will be distributed in the solution. However, surface charge control through checking pH might not be sufficient to create a suspension of minimum agglomeration and maximum stability. Typically, the best practical way to create a stable suspension is using chemical stabilizers called scatters, which are usually polyelectrolyte surfactants that adsorb on particle surface, change its surface charge, and consequently alter the electrostatic forces.

3.5. Fabrication of Nano-Structural Composite Coatings through Electroplating

Different nano-sized particles (in spherical form) in the range of 4 to 800 nm have been successfully deposited in metallic matrixes. Aluminum oxide (Al_2O_3), chromium (Cr), diamond (C), silicon carbide (SiC), gold (Au), Silicon oxide (SiO_2), Zirconium oxide (ZrO_2), Titanium oxide (TiO_2), polystyrene (PS), and silicon nitride (Si_3N_4) are the well-known examples. Among different metals, copper and nickel have been widely used, and alumina particles have been studied more than the other nano-metric particles.

The entrance of nano-metric particles into the coating depends on many process parameters such as particle specifications (concentration, surface charge, type, form, and size), electrolyte composition (electrolyte concentration, additives, temperature, pH, and type of the surfactant), current density (direct current and pulse current), frequency, time percent of illumination, type of bath turbulence (layered, mixed, or turbulent), and electrode geometry (rotating disc, rotating cylinder, parallel sheets of the electrode, and many other conditions).

No research has been conducted on nano-particles of different structures in a similar matrix. Type of the particles (conductive or nonconductive), and their form (nanotubes, nanorods, and nanopills) may have different impacts on particles codeposition. Another factor affecting codeposition process is life of nano-structural particles suspending in the solution and stability of the particle suspension in long time.

3.5.1. Effect of current Type and the Amount of current Density

For creating composite coatings through electroplating, different currents, such as direct, pulse, and reverse pulse currents can be used. Although direct current has been widely used for producing coatings consisting of nano-metric particles, employing pulse current will result in higher presence of nano-metric particles and provides the possibility of producing wider range of composites with their specification. Two or more cathodic direct current, which are repeated in definite periods, are used in pulse current. Pulse reverse current has similar specification, and its only difference is that cathodic current is on during illumination and anodic current is on during darkness. This method have been identified as a successful one for placing higher concentration of nano-metric particles, since a part of the metal placed during darkness is removed through this method.

The reverse pulse current method has been used for creating copper coating containing a high concentration of 30-nanometer alumina particles. Reverse pulse current has three general advantages over direct current: (1) the amount of nano-metric particles can be increased, (2) lower concentration of nano-metric particles can be used in the solution, and (3) selective presence of smaller particles in the coating can be imposed. All these three advantages are resulted from partial dissolution of the metal placed during application of anodic current. Therefore, larger particles tend to separate from the coating, while the smaller ones remain in the coating.

Appropriate and uniform distribution of 80-nanometer alumina particles in nickel coating has been obtained by simultaneous application of ultrasonic and pulse current. Using ultrasonic during deposition, the amount of aluminum whiskers in nickel coating can be reduced. In a frequency of 50 Hz, only large pyramids are observable which are free from aluminum whiskers. In low frequencies, an increase in the amount of aluminum whiskers along with their considerable agglomeration was observed. Using ultrasonic during deposition causes a reduction in particle agglomeration; it also brings about particle separation from the metallic matrix in higher frequencies.

When using both direct and ultrasonic currents, the effect of concentration of bath components on presence of nano-structural particles in the coating has been studied. It has been proved that lower agglomeration rate occur in lower concentrations of metallic ions. For example, in an electrolyte containing 0.2 M Ni^{+2}, average size of the nanostructures is 178 nm. Although low concentration of bath chemical components results in uniform distribution and a lower agglomeration of nano-metric particles, the side reaction of hydrogen release must be considered to avoid current yield reduction and lack of particle attraction to cathode surface due to hydrogen release.

Current density also affects the amount of alumina particles entering nickel coating. Using a sulfamate bath containing 600 to 800 nm particles of α-Al_2O_3, an increase in current density resulted in a rough surface microstructure containing lower amount of aluminum particles. In another study on 32 nm alumina particles, it was found out that particle concentration in the coating depends on type of the electrolyte. In chloride electrolyte, higher presence of nanostructures was observed in low current density, whereas higher concentration of particles in the coating was obtained for citrate electrode in higher current densities. This observation illustrates the fact that the effect of current density depends on electrolyte composition.

3.5.2. Effect of the Size of Nanostructures

Particle size affects particle presence rate in the coating. Previous research has shown that more particles can enter a volume unit of the coating by reducing particle size. For instance, participation rates of alumina particles of two different sizes (50 nm and 300 nm) have been compared in nickel coating. In similar electroplating conditions, it has been proven that volume percentage ratio of 300 nm alumina particles in nickel coating is much higher than that of 50 nm alumina particles. Nonetheless, the number of 50 nm alumina particles in the coating is much larger than that of 300 nm alumina particles.

3.5.3. Effect of pH on Colloidal System Properties and Electroplating of Composite Coatings

3.5.3.1. Zeta Potential

Zeta potential is the electrical potential present at the boundary of solid surface and the liquid medium in which the solid is present; it is also a function of particle surface charge, absorbed layer at common boundary, and type and composition of the medium in which the particle is suspending. Particle zeta potential indicates the particle interaction and can be used as a tool for predicting long-term stability of the solution. When the particles are negatively or positively charge, they tend to repel each other and to resist against agglomeration. Conversely, when the particles have low zeta potential (near to 0), there will not be any factor preventing particle agglomeration; and the particles will therefore become agglomerated. The boundary between the stable and unstable solutions is +20 mV and -20 mV. It means that if zeta potential of the solution is higher than +20 mV or lower than -20 mV, the solution is usually stable; and the solution is unstable when the zeta potential is between +20 mV and -20 mV.

Most materials acquire zeta potential when suspending in water. Zeta potential depends on different factors such as pH, additive concentration, and ionic strength of the solution. For many chemically simple surfaces, e.g. mineral oxides placed in dilute solution of salt, zeta potential shows a definite behavior by pH change: when pH increases by adding basic materials, zeta potential becomes more negative or at least its positive amount reduces. When adding an acid to the water, acid ionization will reduce OH^- ions and create a more positive charge on particle surface. Accordingly, mineral oxides in dilute salt solutions are lowly positive in low pHs and highly negative in high pHs. In diagrams presenting zeta potential against pH, the curve passes the zero potential point in some pHs, which is called isoelectric point (iep) and is practically significant since it indicates the pH in which the solution has minimum stability.

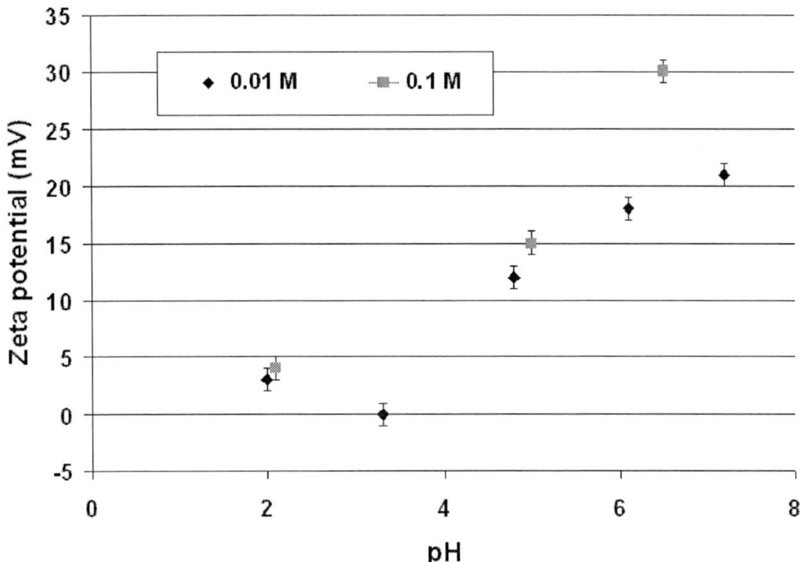

Figure 3.1. Variation of zeta potential versus pH in different concentrations of nickel sulfamate solution.

As mentioned earlier, the medium in which the particle is placed affects particle zeta potential. For instance, as can bee seen in figure 3.1, zeta potential of 62 nm silicon carbide particles increase correspondingly to the increase in amount of nickel sulfamate in electroplating solution. Generally, silicon carbide particles are negatively charged; when these particles are place in nickel electroplating solution, they attract nickel ions on their surface and acquire a positive surface charge. Because of the increase in nickel sulfamate in electroplating bath, density of the nickel ions adsorbed on the particles increases; consequently, this will cause a rise in zeta potential of the particles.

3.5.3.2. How to Determine Zeta Potential

The principles of zeta potential determination are very simple. A controlled electric field is created between the electrodes placed in the suspension, which cause the charged particles to move toward the electrode having opposite polarity. Viscous force imposed on moving particles tends to impede their displacement. Therefore, a speed is created which is balanced between the electrostatic attraction and viscosity. Zeta potential is obtained from measuring mobility distribution of suspending charged particles within an electric filed. Depending on ions concentration in a solution, Smoluchowski relation can be used for solutions having high ionic power, and Huckel relation for solutions having low ionic power to determine zeta potential according to particles mobility rate.

3.5.3.3. Effect of pH on Properties of the Colloidal System

When the particles suspend in the solution, their surface charge can cause them to scatter or agglomerate. By changing pH, the particles can be agglomerated or suspended depending on their polarity and surface charge. Figure 3.2 indicates dependency of particle size on pH in the suspension containing 5 weight percent of alumina (0.6μm) when not using any scatter.

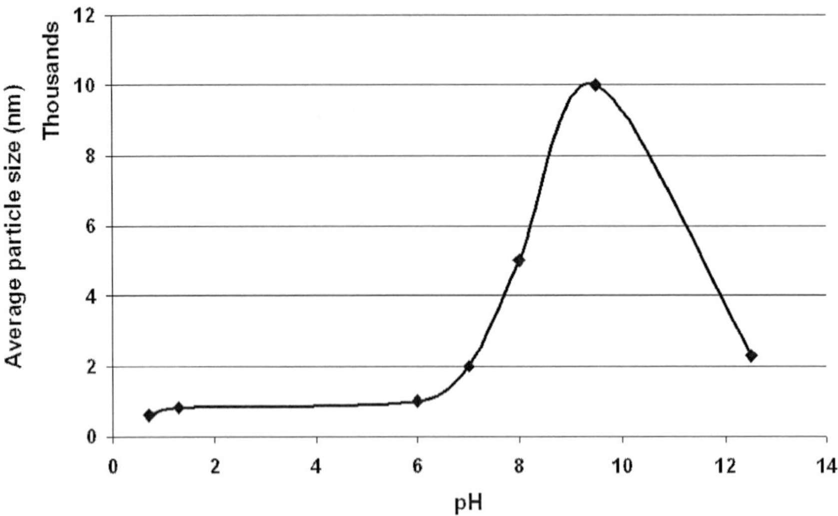

Figure 3.2. pH effect on size distribution of alumina particles without using surfactants.

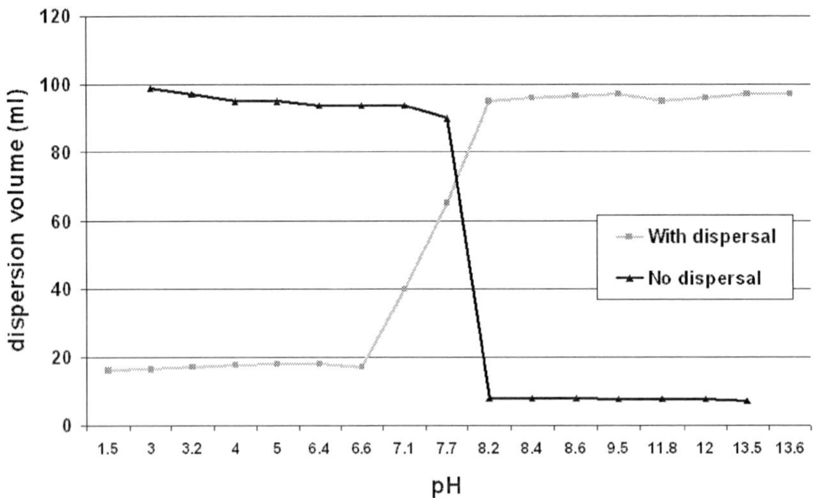

Figure 3.3. pH effect on dispersion volume.

By increasing pH, the size of agglomerated particles become larger until it reaches a maximum around pH = 9.5 which is very close to iep. The increase in agglomerates size is because of the decrease in absolute value of zeta potential. Therefore, repulsion between the particles is decreased. The effect of the increase in agglomerates size reveals itself in suspension volume and solution viscosity: as particles enlarge, suspension volume is reduced and its viscosity is increased.

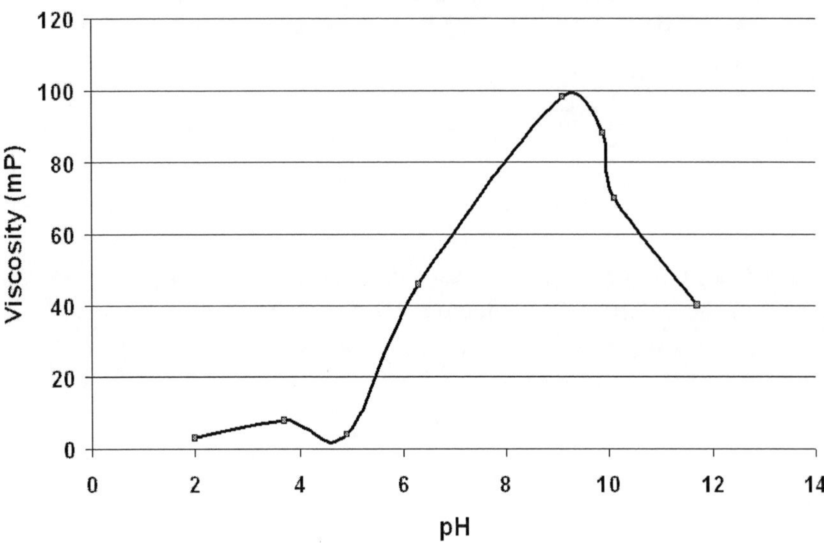

Figure 3.4. change of solution viscosity by pH.

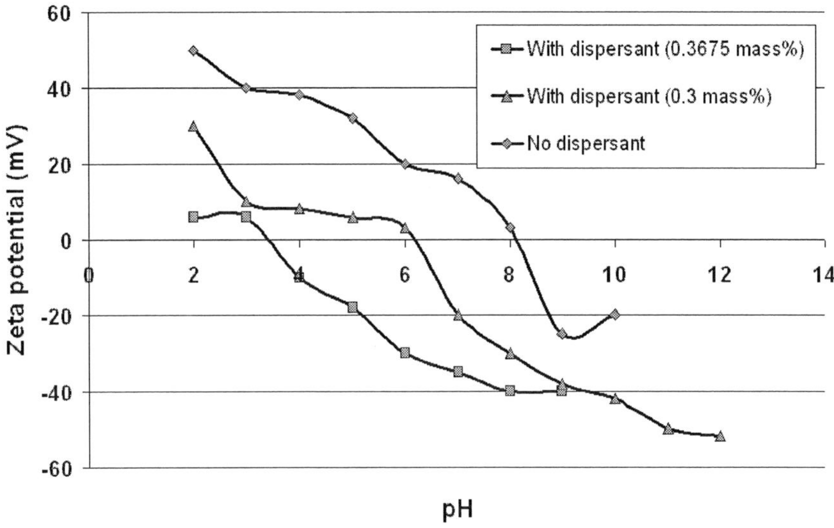

Figure 3.5. dependence of zeta potential on pH and additive concentration.

Figure 3.5 presents the effect of scatter APC on zeta potential of the particles; as can bee seen in this figure, presence of the scatter cause a displacement in iep toward acidic values (pH = 3.6). In fact, carboxyl groups released from the scatter are negatively charged and are absorbed on positively-charged alumina particles to induce a negative surface charge on them.

Considering the aforementioned points, we can conclude that by changing pH and adjusting it at optimum value, one can prevent particle agglomeration in colloidal systems used in electroplating. In addition, particle surface potential can be increased to create a better condition for particles to participate in the coating.

3.5.4. Surfactant Effect on Colloidal System Properties and Composite Coating Electroplating

Surfactants are capable of being adsorbed on particle surfaces to increase zeta potential of the particles. In this way, they prevent particle agglomeration and stabilize the colloidal system. Moreover, they can result in higher participation of the particles in the coating through creating a positive zeta potential on the particles. Some surfactants adsorb on particle surface and reduce as the particle reaches the cathode surface; they accordingly facilitate particle burial in the coating. Another group of surfactants act as catalyzing for reducing metallic ions and cause quick participation of the particles into the metallic matrix.

In a study with the aim of increasing volume of alumina particles in nickel coating, the researchers have used hexadecyle pyridinium bromide (HPB). The initial size of the alumina powder is 0.8 μm. Figure 3.6 indicates the concentration effect of the surfactant used in this study on volume percent on alumina participated in the coating. As is observable in the figure, increasing surfactant concentration to 100 mg will be of no effect on the alumina volume in the coating. However, after 100 mg, an elevating trend is seen until the surfactant is again of weak effect on increasing alumina volume in the coating.

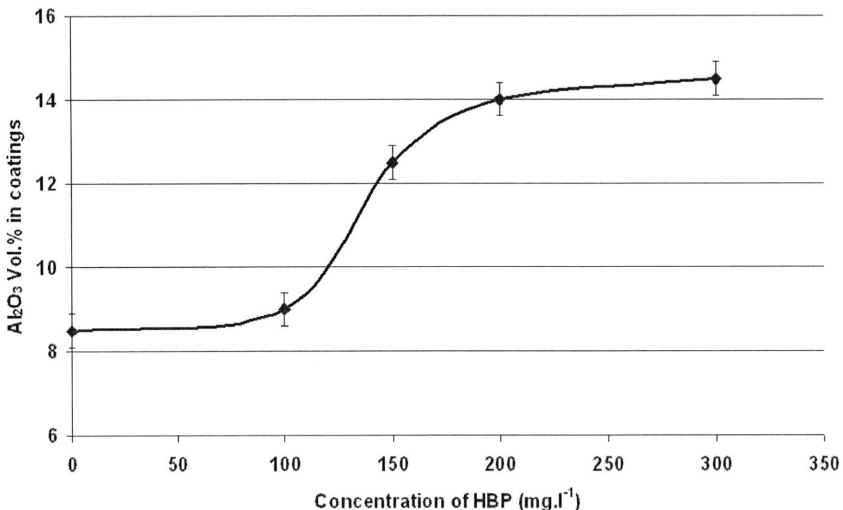

Figure 3.6. volume fraction of alumina codeposited particles in different concentrations of HPB.

As is inferable from figure 3.7 (presenting surfactant effect on zeta potential), adding more surfactant will increase zeta potential of the particles so that particles zeta potential will rich +20 mV at 300 mg/l of HPB.

Based on Guglielmi's theory, conversion of weak particle attraction to a strong attraction during an electrochemical process controls the rate of particle participation in the coating. The positive potential generated on the particles both facilitates particles movement toward the cathode and causes strong adsorption of the particles to the cathode. In another way, using surfactants increases the intensity of both weak and strong particle attractions. In the case of HPB, adding more than 150mg/l of it to the bath will be of no considerable effect on rise of volume percentage of the particles; the reason for this observations is the impulsion increase occurring between the surfactant layer absorbed on the cathode and the particles approaching the cathode. Furthermore, particles distribution mode in the solution affects their distribution mode in the coating. When adding adequate amount of HPB to the solution, alumina particles are scattered in the solution and create a more stable solution.

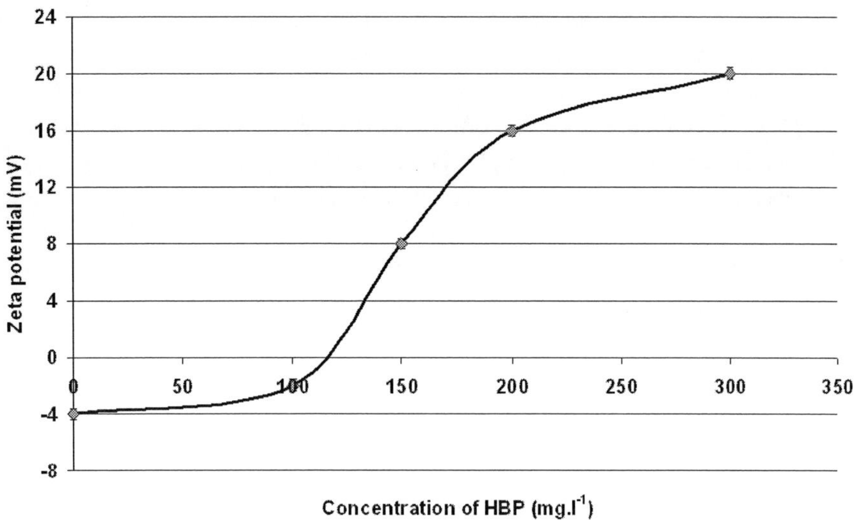

Figure 3.7. zeta potential of the suspending particles of alumina varying in electroplating solution by HPB concentration.

3.6. Effect of Silicon Carbide Particles on Deposition of Ni-SiC Composite Coating

In the beginning of deposition process, nickel ions are adsorbed on the cathode and react with hydroxide ions through following equations and are converted to single nickel atoms in two steps. Then, the single nickel atoms are placed in suitable locations and create a deposition layer.

$$Ni^{2+} + OH^{-} \rightarrow Ni(OH)^{+}$$

$$Ni(OH)^{+} \rightarrow Ni(OH)^{+}_{ads}$$

$$Ni(OH)^{+}_{ads} + 2e^{-} \rightarrow Ni + OH^{-}$$

Two first equations are considered as key steps determining reaction rate and are used when writing the equations of the corresponding circuit. When silicon carbide particles are present, attractions and impulsions occurring on the cathode surface have different time constants. Nickel cathode potentiodynamic diagrams in both absence and presence of 20 nm silicon carbide particles in the bath, and electrode rotation speed equal to 200 rpm were obtained. Adding carbide silicon nano-particles displaces nickel reduction curve toward more positive potentials. The change in reduction potential is attributed to the increase in cathode active surface because of increasing surface roughness by presence of silicon carbide particles and to possible increase in ions movement by nano-particles.

Impedance curves were drawn in -750 mV cathodic potential in both absence and presence of silicon carbide nano-particles in the bath, and in two electrode rotation speed of 100 and 200 rpm. In low frequencies induction characteristic semicircles of pure nickel is larger than the ones for nickel coating containing silicon carbide nanostructures. Resistance against charge transfer under absence or presence of silicon carbide nano-metric particles is found to be 6.5 ohm/cm^2 and 8.2 ohm/cm^2 respectively (in reduction potential of -750mV compared with the reference electrode (Ag/AgCl), and electrode rotation speed of 200 rpm). Obviously, resistance against charge transfer is lower under presence of silicon carbide nano-metric particles. In addition, the effect of nano-particles on impedance diagrams during electroplating is more obvious when using lower cathode rotation speeds. This observation proves the hypothesis that nano-particles facilitate ions displacement toward the cathode by transferring the ions adsorbed on them. In lower cathode rotation speeds, in which nickel electroplating is controlled by mass transfer, the effect of nano-particles on impedance diagram is more noticeable.

3.7. Characteristics of Nano-Composite Coatings

The main objectives of producing nano-composite coatings are to increase their hardness and resistance against wear and corrosion. Several terms must be in mind for studying nanocomposites: (1) nano-crystalline structures have higher micro-hardness compared with microcrystal structures; (2) the coatings obtained from a pulse current are of higher hardness compared with the ones obtained from a direct current.

Silicon carbide-nickel composite coatings were found to have more hardness compared with pure nickel coatings. Some researchers argue that silicon carbide particles act as nuclei center and, therefore, prevent growth of the grains; some other researchers believe that SiC particles adsorb hydrogen ions and create a temporary basic condition in the coating-electrolyte common region; they finally result in presence of the structure (211) having higher solidity compared with the more ductile structure. In fact, the ductile structure (100) is the dominant structure of the coating under absence of silicon carbide in the solution, which is a structure with high ductility, low hardness, and high internal stress. Furthermore, this group of researchers believes that the temporary change in electrolyte composition prevents growth of the grains, induces nucleation, and consequently helps increase coating hardness by creating a more fine-grained structure.

Another factor affecting coating hardness might be the effect of scattering-hardness mechanism. In other words, the particles placed in the coating act as obstacles of dislocations. It is worth noting that this effect can be used best when a uniform distribution of non-agglomerated particles (smaller than 1μm) is created in the coating.

It has been reported that the reduction in size of silicon carbide particles positively affects resistance of silicon carbide-nickel composite coatings against wear and corrosion. Nonetheless, the reduction in particle size causes a decrease in rate of particle presence in the coating. Hardness has been also proved to be totally dependent on particle size, and the coating containing nano-metric SiC particle presents higher hardness compared with its rival when volume percent of the nano-metric particles present in the coating is lower than volume percent of micrometric particles present in the coating containing silicon-carbide micrometric particles. This fact indicates that not also volume percent of the particles participated in the coating is important, but density of them must also be considered.

Moreover, some researchers believe that a change in hardness of the coatings containing micrometric and nano-metric particles might occur as a result of different codeposition mechanisms which are dependent on particle size. To be more specific, some of the nano-metric particles are placed at grain boundaries and some inside the grains, whereas all the micrometric particles are placed at grain boundaries and margins.

Corrosion rate of a nickel coating composed of titanium oxide in an acidic solution (5% NaCl, 1g/dm^3 $CuCl_2.2H_2O$, 50°C, 1 week immersion) was evaluated and it was observed that the corrosion rate depends on titanium oxide particle size and number in the nickel coating. In addition, presence of titanium oxide nano-sized particles in the nickel coating increases resistance of the nickel coating against corrosion compared with the pure nickel. A coating composed of Anatase nano-sized particles (12 nm in size) had lower corrosion rate compared with a coating composed of 1μm Rutile particles. However, the results obtained here can not be generalized to all the composite coatings. For example, in nickel coatings composed of 150 nm alumina particles and 300 nm titanium oxide particles, corrosion rate is higher than the coatings without particles. Wear resistance has been compared between the nickel coatings containing 10 to 30 nm zirconium particles and produced by direct, pulse, and reverse current. Wear test was conducted using abrasive balls on a disk and under dry sliding condition at 160 rpm, a load of 11.3 N, and 1200 cycles. The highest hardness was that of the coating obtained form reverse pulse current, and lowest hardness was obtained for direct current. All the coatings acquired had better resistance against wear compared with pure nickel.

3.8. Example of Characterization of Nanocomposite Coatings

In the case of polymeric-based nanocomposite layers, hard binary silica/epoxy nanocomposite coatings were successfully prepared. Wear and electrical conductivity tests have been done on the coatings. Increasing the coating time will lead to slight increase in density of different nanoparticulates, decreasing wear rate and increasing the electrical conductivity of obtained layers. Obtained relations for density of different nanoparticulates, wear rate and electrical conductivity is linear, quadratic and linear respectively. AFM nanostructures illustrated that planar nanostructure of coating in low coating times will change to multidirectional nanostructure in high coating times. Increasing the rotating speed will lead to decrease in density of different nanoparticulates, increasing wear rate and decreasing the electrical conductivity of obtained layers. Type of obtained relations is as same as for the effect of coating time. Difference of electrical conductivity is due to the amount of conductive nanoparticulates in the coating. Wear rate illustrates an optimum level and increasing the SiO_2/QD ratio after this level will decrease wear rate significantly. Electrical conductivity changes against SiO_2/QD ratio can be divided in to three sections. Slight increase follows by a severe increase in second section and the third section is like the first one (all of them can assume linear). Effective share of QD nanoparticulates in electrical conductivity of obtained layers, their amount in coating and bridging among them will cause the rapid increase in second section of the curve.

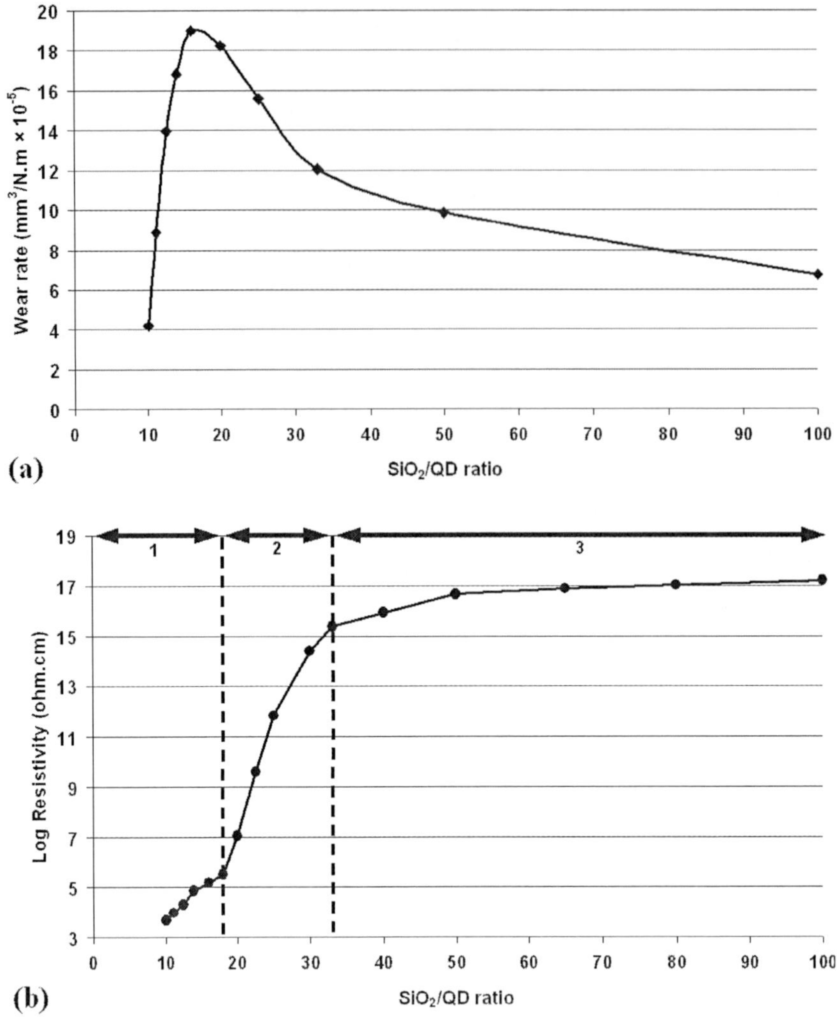

Figure 3.8. Effect of SiO_2/QD ratio on (a) wear rate (b) electrical resistivity of different obtained layers.

References

[1] Aliofkhazraei, M., Sabour Rouhaghdam, A. and Heydarzadeh, A. Strong relation between corrosion resistance and nanostructure of compound layer of treated 316 austenitic stainless steel. *Materials Characterization,* 2009, 60(2), 83-89.

[2] Aliov, M.K. and Sabur, A.R. Formation of a novel hard binary SiO_2/quantum dot nanocomposite with predictable electrical conductivity. *Modern Physics Letters* B, 2010, 24(1), 89-96.

[3] Amerio, E., Sangermano, M., Malucelli, G., Priola, A. and Rizza, G. Preparation and characterization of hyperbranched polymer/silica hybrid nanocoatings by dual-curing process. *Macromolecular Materials and Engineering,* 2006, 291(10), 1287-1292.

[4] Amerio, E., Sangermano, M., Malucelli, G., Priola, A. and Voit, B. Preparation and characterization of hybrid nanocomposite coatings by photopolymerization and sol-gel process. *Polymer*, 2005, 46(25), 11241-11246.

[5] Aouadi, S.M., Wong, K.C., Mitchell, K.A.R., Namavar, F., Tobin, E., Mihut, D.M. and Rohde, S.L. Characterization of titanium chromium nitride nanocomposite protective coatings. *Applied Surface Science*, 2004, 229(1-4), 387-394.

[6] Audronis, M., Leyland, A., Matthews, A., Wen, J.G. and Petrov, I. Characterization studies of pulse magnetron sputtered hard ceramic titanium diboride coatings alloyed with silicon. *Acta Materialia*, 2008, 56(16), 4172-4182.

[7] Avramova, I., Stefanov, P., Nicolova, D., Stoychev, D. and Marinova, T. Characterization of nanocomposite CeO_2-Al_2O_3 coatings electrodeposited on stainless steel. *Composites Science and Technology*, 2005, 65(11-12), 1663-1667.

[8] Azami, M., Moztarzadeh, F. and Tahriri, M. Preparation, characterization and mechanical properties of controlled porous gelatin/hydroxyapatite nanocomposite through layer solvent casting combined with freeze-drying and lamination techniques. *Journal of Porous Materials*, 2009, 1-8.

[9] Barshilia, H.C., Deepthi, B. and Rajam, K.S. Deposition and characterization of TiAlN/Si_3N_4 superhard nanocomposite coatings prepared by reactive direct current unbalanced magnetron sputtering. *Vacuum*, 2006, 81(4), 479-488.

[10] Barshilia, H.C., Deepthi, B. and Rajam, K.S. Deposition and characterization of CrN/Si_3N_4 and CrAlN/Si_3N_4 nanocomposite coatings prepared using reactive DC unbalanced magnetron sputtering. *Surface and Coatings Technology*, 2007, 201(24), 9468-9475.

[11] Bauer, F., Ernst, H., Hirsch, D., Naumov, S., Pelzing, M., Sauerland, V. and Mehnert, R. Preparation of scratch and abrasion resistant polymeric nanocomposites by monomer grafting onto nanoparticles, 5a: Application of mass spectroscopy and atomic force microscopy to the characterization of silane-modified silica surface. *Macromolecular Chemistry and Physics*, 2004, 205(12), 1587-1593.

[12] Cai, Y., Wu, N., Wei, Q., Zhang, K., Xu, Q., Gao, W., Song, L. and Hu, Y. Structure, surface morphology, thermal and flammability characterizations of polyamide6/organic-modified Fe-montmorillonite nanocomposite fibers functionalized by sputter coating of silicon. *Surface and Coatings Technology*, 2008, 203(3-4), 264-270.

[13] Castagno, K.R.L., Dalmoro, V., Mauler, R.S. and Azambuja, D.S. Characterization and corrosion protection properties of polypyrrole / montmorillonite electropolymerized onto aluminium alloy 1100. *Journal of Polymer Research,* 2009, 1-9.

[14] Chang, C.F., Wu, Y.L. and Hou, S.S. Preparation and characterization of superparamagnetic nanocomposites of aluminosilicate/silica/magnetite. *Colloids and Surfaces A: Physicochemical and Engineering Aspects,* 2009, 336(1-3), 159-166.

[15] Chang, Y.Y., Yang, S.J. and Wang, D.Y. Characterization of TiCr(C,N)/amorphous carbon coatings synthesized by a cathodic arc deposition process. *Thin Solid Films*, 2007, 515(11), 4722-4726.

[16] Chen, A., Huang, X., Tong, Z., Bai, S., Luo, R. and Liu, C.C. Preparation, characterization and gas-sensing properties of SnO_2-In_2O_3 nanocomposite oxides. *Sensors and Actuators, B: Chemical*, 2006, 115(1), 316-321.

[17] Chen, C.H., Li, H.Y., Chien, C.Y., Yen, F.S., Chen, H.Y. and Lin, J.M. Preparation and characterization of Al_2O_3/Nylon 6 nanocomposite masterbatches. *Journal of Applied Polymer Science*, 2009, 112(2), 1063-1069.
[18] Chen, K., Wang, C., Huang, Y. and Lin, W. Preparation and characterization of polymer-clay nanocomposite films. *Science in China,* Series B: Chemistry, 2009, 52(12), 2323-2328.
[19] Chen, Q.Z. and Boccaccini, A.R. Poly(D,L-lactic acid) coated 45S5 Bioglass®-based scaffolds: Processing and characterization. *Journal of Biomedical Materials Research -* Part A, 2006, 77(3), 445-457.
[20] Chen, S., You, B., Zhou, S. and Wu, L. Preparation and characterization of scratch and mar resistant waterborne epoxy/silica nanocomposite clearcoat. *Journal of Applied Polymer Science*, 2009, 112(6), 3634-3639.
[21] Chen, X. Synthesis and characterization of ATO/SiO_2 nanocomposite coating obtained by sol-gel method. *Materials Letters*, 2005, 59(10), 1239-1242.
[22] Chen, Y., Zhou, S., Yang, H., Gu, G. and Wu, L. Preparation and characterization of nanocomposite polyurethane. *Journal of Colloid and Interface Science*, 2004, 279(2), 370-378.
[23] Chiang, C.L. and Chang, R.C. Synthesis, characterization and properties of novel self-extinguishing organic-inorganic nanocomposites containing nitrogen, silicon and phosphorus via sol-gel method. *Composites Science and Technology*, 2008, 68(14), 2849-2857.
[24] Cioffi, N., Ditaranto, N., Torsi, L., Picca, R.A., Sabbatini, L., Valentini, A., Novello, L., Tantillo, G., Bleve-Zacheo, T. and Zambonin, P.G. Analytical characterization of bioactive fluoropolymer ultra-thin coatings modified by copper nanoparticles. *Analytical and Bioanalytical Chemistry*, 2005, 381(3), 607-616.
[25] Cranston, E.D. and Gray, D.G. Morphological and optical characterization of polyelectrolyte multilayers incorporating nanocrystalline cellulose. *Biomacromolecules*, 2006, 7(9), 2522-2530.
[26] Darbandi, A.J., Enz, T. and Hahn, H. Synthesis and characterization of nanoparticulate films for intermediate temperature solid oxide fuel cells. *Solid State Ionics*, 2009, 180(4-5), 424-430.
[27] Delozier, D.M., Watson, K.A., Smith, J.G. and Connell, J.W. Preparation and characterization of space durable polymer nanocomposite films. *Composites Science and Technology*, 2005, 65(5 SPEC. ISS.), 749-755.
[28] Devaux, E., Koncar, V., Kim, B., Campagne, C., Roux, C., Rochery, M. and Saihi, D. Processing and characterization of conductive yarns by coating or bulk treatment for smart textile applications. *Transactions of the Institute of Measurement and Control,* 2007, 29(3-4), 355-376.
[29] Dinh, N.N., Chi, L.H., Long, N.T., Thuy, T.T.C., Trung, T.Q. and Kim, H.K. Preparation and characterization of nanostructured composite films for organic light emitting diodes. *Journal of Physics: Conference Series*, 2009, 187.
[30] Dong, J., Xu, Z. and Wang, F. Engineering and characterization of mesoporous silica-coated magnetic particles for mercury removal from industrial effluents. *Applied Surface Science*, 2008, 254(11), 3522-3530.

[31] Enomoto, H., Matsumoto, S. and Lerner, M.M. Synthesis and characterization of conductive polypyrrole/montmorillonite nanocomposite. *Japanese Journal of Applied Physics*, Part 2: Letters, 2005, 44(1-7), L224-L226.

[32] Feng, X., Huang, H., Ye, Q., Zhu, J.J. and Hou, W. Ag/polypyrrole core-shell nanostructures: Interface polymerization, characterization, and modification by gold nanoparticles. *Journal of Physical Chemistry* C, 2007, 111(24), 8463-8468.

[33] Fuke, M.V., Adhyapak, P.V., Mulik, U.P., Amalnerkar, D.P. and Aiyer, R.C. Electrical and humidity characterization of m-NA doped Au/PVA nanocomposites. *Talanta*, 2009, 78(2), 590-595.

[34] Galvan, D., Pei, Y.T. and De Hosson, J.T.M. TEM characterization of a Cr/Ti/TiC graded interlayer for magnetron-sputtered TiC/a-C:H nanocomposite coatings. *Acta Materialia*, 2005, 53(14), 3925-3934.

[35] Galya, T., SedlarŒik, V., KurŒitka, I., Novotny, R., SedlarŒikova, J. and Saha, P. Antibacterial poly(vinyl alcohol) film containing silver nanoparticles: preparation and characterization. *Journal of Applied Polymer Science*, 2008, 110(5), 3178-3185.

[36] Ge, C., Xie, C., Zeng, D. and Cai, S. Formaldehyde-, benzene-, and xylene-sensing characterizations of Zn-W-O nanocomposite ceramics. *Journal of the American Ceramic Society*, 2007, 90(10), 3263-3267.

[37] Geblinger, N., Thiruvengadathan, R. and Regev, O. Preparation and characterization of a double filler polymeric nanocomposite. *Composites Science and Technology,* 2007, 67(5), 895-899.

[38] Golub, A.S., Payen, C., Protzenko, G.A., Novikov, Y.N. and Danot, M. Nanocomposite materials consisting of alternating layers of molybdenum disulfide and cobalt or nickel hydroxides: Magnetic characterization. *Solid State Communications*, 1997, 102(5), 419-423.

[39] Gorokhovsky, V., Bowman, C., Gannon, P., VanVorous, D., Voevodin, A.A. and Rutkowski, A. Tribological performance of hybrid filtered arc-magnetron coatings. Part II: Tribological properties characterization. *Surface and Coatings Technology*, 2007, 201(14), 6228-6238.

[40] Gorokhovsky, V., Bowman, C., Gannon, P., VanVorous, D., Voevodin, A.A., Rutkowski, A., Muratore, C., Smith, R.J., Kayani, A., Gelles, D., Shutthanandan, V. and Trusov, B.G. Tribological performance of hybrid filtered arc-magnetron coatings. Part I: Coating deposition process and basic coating properties characterization. *Surface and Coatings Technology,* 2006, 201(6), 3732-3747.

[41] Gorokovsky, V.I., Bowman, C., Gannon, P.E., VanVorous, D., Voevodin, A.A., Muratore, C., Kang, Y.S. and Hu, J.J. Deposition and characterization of hybrid filtered arc/magnetron multilayer nanocomposite cermet coatings for advanced tribological applications. *Wear*, 2008, 265(5-6), 741-755.

[42] Grandfield, K. and Zhitomirsky, I. Electrophoretic deposition of composite hydroxyapatite-silica-chitosan coatings. *Materials Characterization*, 2008, 59(1), 61-67.

[43] Ha, E.H., Huang, D.Q., Ha, E.P., Wang, Z.Y. and Ding, H.Y. Preparation and characterization of carbon nanotube/conducting polymer nanocomposites. *Cailiao Gongcheng/Journal of Materials Engineering*, 2008(10), 122-125.

[44] Ha, Y.G., You, E.A., Kim, B.J. and Choi, J.H. Fabrication and characterization of OLEDs using MEH-PPV and SWCNT nanocomposites. *Synthetic Metals*, 2005, 153(1-3), 205-208.

[45] Habibi, M.H. and Nasr-Esfahani, M. Preparation, characterization and photocatalytic activity of a novel nanostructure composite film derived from nanopowder TiO_2 and sol-gel process using organic dispersant. *Dyes and Pigments*, 2007, 75(3), 714-722.

[46] Habibi, M.H., Nasr-Esfahani, M. and Egerton, T.A. Preparation, characterization and photocatalytic activity of TiO_2 / Methylcellulose nanocomposite films derived from nanopowder TiO_2 and modified sol-gel titania. *Journal of Materials Science*, 2007, 42(15), 6027-6035.

[47] He, L.P., Mai, Y.W. and Chen, Z.Z. Fabrication and characterization of nanometer CaP(aggregate)/Al_2O_3 composite coating on titanium. *Materials Science and Engineering* A, 2004, 367(1-2), 51-56.

[48] He, X., Xu, S., Sklyarov, A.V. and Hardcastle, S. Thin-film synthesis and microstructure characterization of the poly(vinyl alcohol) matrix with functionalized carbon nanotubes. *Journal of Materials Research*, 2007, 22(7), 1872-1878.

[49] Hong, R.Y., Li, J.H., Qu, J.M., Chen, L.L. and Li, H.Z. Preparation and characterization of magnetite/dextran nanocomposite used as a precursor of magnetic fluid. *Chemical Engineering Journal*, 2009, 150(2-3), 572-580.

[50] Huang, Z., Guan, J. and Wang, Y. Preparation and characterization of glass sphere/silver core-shell nanocomposite particles. Kuei Suan Jen Hsueh Pao/ *Journal of the Chinese Ceramic Society*, 2006, 34(7), 882-886.

[51] Jee, A.Y. and Lee, M. Surface functionalization and physicochemical characterization of diamond nanoparticles. *Current Applied Physics,* 2009, 9(2 SUPPL.), e144-e147.

[52] Jeevananda, T., Siddaramaiah, Kim, N.H., Heo, S.B. and Lee, J.H. Synthesis and characterization of polyaniline-multiwalled carbon nanotube nanocomposites in the presense of sodium dodecyl sulfate. *Polymers for Advanced Technologies*, 2008, 19(12), 1754-1762.

[53] Jeon, J.H., Choi, S.R., Chung, W.S. and Kim, K.H. Synthesis and characterization of quaternary Ti-Si-C-N coatings prepared by a hybrid deposition technique. *Surface and Coatings Technology*, 2004, 188-189(1-3 SPEC.ISS.), 415-419.

[54] Jiang, J. Facile synthesis and characterization of polyaniline/$NiFe_2O_4$ nanocomposite in w/o microemulsion. *Journal of Macromolecular Science,* Part B: Physics, 2008, 47(2), 242-249.

[55] Jiang, J. and Ai, L.H. Microemulsion-mediated in-situ synthesis and magnetic characterization of polyaniline/$Zn_{0.5}Cu_{0.5}Fe_2O_4$ nanocomposite. *Applied Physics A: Materials Science and Processing*, 2008, 92(2), 341-344.

[56] Jiang, J., Chen, C., Ai, L.H., Li, L.C. and Liu, H. Synthesis and characterization of novel ferromagnetic PPy-based nanocomposite. *Materials Letters*, 2009, 63(5), 560-562.

[57] Jiang, L. and Gao, L. Fabrication and characterization of ZnO-coated multi-walled carbon nanotubes with enhanced photocatalytic activity. *Materials Chemistry and Physics*, 2005, 91(2-3), 313-316.

[58] Kamalan Kirubaharan, A.M., Selvaraj, M., Maruthan, K. and Jeyakumar, D. Synthesis and characterization of nanosized titanium dioxide and silicon dioxide for corrosion resistance applications. *Journal of Coatings Technology Research*, 2009, 1-8.

[59] Kazmanli, K., Daryal, B. and Urgen, M. Characterization of nano-composite TiN-Sb coating produced with hybrid physical vapor deposition system. *Thin Solid Films*, 2007, 515(7-8), 3675-3680.

[60] Khazrayie, M.A. and Aghdam, A.S.R. Characterization of Ni-W/MWCNT nanocomposite layers formed by pulsed electrochemical deposition. *Protection of Metals*, 2010(6).

[61] Kota, A.K., Cipriano, B.H., Powell, D., Raghavan, S.R. and Bruck, H.A. Quantitative characterization of the formation of an interpenetrating phase composite in polystyrene from the percolation of multiwalled carbon nanotubes. *Nanotechnology,* 2007, 18(50).

[62] Kuila, B.K., Garai, A. and Nandi, A.K. Synthesis, optical, and electrical characterization of organically soluble silver nanoparticles and their poly(3-hexylthiophene) nanocomposites: Enhanced luminescence property in the nanocomposite thin films. *Chemistry of Materials*, 2007, 19(22), 5443-5452.

[63] Kulisch, W., Colpo, P., Rossi, F., Shtansky, D.V. and Levashov, E.A. Characterization of a hybrid PVD/PACVD system for the deposition of TiC/CaO nanocomposite films by OES and probe measurements. *Surface and Coatings Technology*, 2004, 188-189(1-3 SPEC.ISS.), 714-720.

[64] Kvien, I., Tanem, B.S. and Oksman, K. Characterization of cellulose whiskers and their nanocomposites by atomic force and electron microscopy. *Biomacromolecules,* 2005, 6(6), 3160-3165.

[65] Landry, V., Riedl, B. and Blanchet, P. Alumina and zirconia acrylate nanocomposites coatings for wood flooring: Photocalorimetric characterization. *Progress in Organic Coatings*, 2008, 61(1), 76-82.

[66] Lang, S., Beck, T., Schattke, A., Uhlaq, C. and Dinia, A. Characterization of nanostructured coatings based on oxides for tribological applications. *Surface and Coatings Technology*, 2004, 180-181, 85-89.

[67] Lee, M.H., Chung, W.J., Park, S.K., Kim, M.S., Seo, H.S. and Ju, J.J. Structural and optical characterizations of multi-layered and multi-stacked PbSe quantum dots. *Nanotechnology,* 2005, 16(8), 1148-1152.

[68] Li, F., Zhou, S., Gu, G., You, B. and Wu, L. Preparation and characterization of ultraviolet-curable nanocomposite coatings initiated by benzophenone/n-methyl diethanolamine. *Journal of Applied Polymer Science*, 2005, 96(3), 912-918.

[69] Li, F., Zhou, S. and Wu, L. Preparation and characterization of UV-curable MPS-modified silica nanocomposite coats. *Journal of Applied Polymer Science,* 2005, 98(5), 2274-2281.

[70] Li, Y.F., Ouyang, J.H., Zhong, J.Y. and Zhou, Y. Preparation and characterization of Al_2O_3-coated $SrSO_4$ nanocomposite powders. Rengong Jingti Xuebao/*Journal of Synthetic Crystals*, 2009, 38(SUPPL. 1), 21-24.

[71] Lin, J., Moore, J.J., Mishra, B., Pinkas, M. and Sproul, W.D. Syntheses and characterization of TiC/a:C composite coatings using pulsed closed field unbalanced magnetron sputtering (P-CFUBMS). *Thin Solid Films*, 2008, 517(3), 1131-1135.

[72] Liu, X., Wang, J., Zhang, J., Liu, B., Zhou, J. and Yang, S. Fabrication and characterization of Ag/polymer nanocomposite films through layer-by-layer self-assembly technique. *Thin Solid Films*, 2007, 515(20-21), 7870-7875.

[73] Lu, C., Donch, I., Nolte, M. and Fery, A. Au nanoparticle-based multilayer ultrathin films with covalently linked nanostructures: Spraying layer-by-layer assembly and

mechanical property characterization. *Chemistry of Materials*, 2006, 18(26), 6204-6210.

[74] Lv, W., Luo, Z., Weng, W. and Yang, H. Synthesis, characterization and performance of nanosized Bi 0.1Sn0.9O2-WPU organic-inorganic hybrid coatings. *Journal of Sol-Gel Science and Technology*, 2009, 51(1), 58-62.

[75] Ma, D., Ma, S. and Xu, K. The tribological and structural characterization of nano-structured Ti-Si-N films coated by pulsed-d.c. plasma enhanced CVD. *Vacuum*, 2005, 79(1-2), 7-13.

[76] Ma, S., Ma, D., Xu, K. and Jie, W. Structural characterization of nanocomposite Ti-Si-N coatings prepared by pulsed dc plasma-enhanced chemical vapor deposition. *Journal of Vacuum Science and Technology B: Microelectronics and Nanometer Structures*, 2004, 22(4), 1694-1698.

[77] Ma, S.L., Ma, D.Y., Guo, Y., Xu, B., Wu, G.Z., Xu, K.W. and Chu, P.K. Synthesis and characterization of super hard, self-lubricating Ti-Si-C-N nanocomposite coatings. *Acta Materialia*, 2007, 55(18), 6350-6355.

[78] Madbouly, S.A. and Otaigbe, J.U. Recent advances in synthesis, characterization and rheological properties of polyurethanes and POSS/polyurethane nanocomposites dispersions and films. *Progress in Polymer Science* (Oxford), 2009, 34(12), 1283-1332.

[79] Madhavan, K., Gnanasekaran, D. and Reddy, B.S.R. Synthesis and characterization of poly(dimethylsiloxaneurethane) nanocomposites: Effect of (in)completely condensed silsesquioxanes on thermal, morphological, and mechanical properties. *Journal of Applied Polymer Science,* 2009, 114(6), 3659-3667.

[80] Mammeri, F., Le Bourhis, E., Rozes, L. and Sanchez, C. Elaboration and mechanical characterization of nanocomposites thin films. Part I: Determination of the mechanical properties of thin films prepared by in situ polymerisation of tetraethoxysilane in poly(methyl methacrylate). *Journal of the European Ceramic Society*, 2006, 26(3), 259-266.

[81] Marino, I.G., Lottici, P.P., Rozzetti, C., Montenero, A., Rocchetti, M., Toselli, M., Marini, M. and Pilati, F. Polariscopic imaging and vibrational characterization of hybrid films for packaging. *Packaging Technology and Science*, 2008, 21(6), 329-338.

[82] Mavis, B., Demirta, T.T., Gmdereliolu, M., Gundez, G. and olak, U. Synthesis, characterization and osteoblastic activity of polycaprolactone nanofibers coated with biomimetic calcium phosphate. *Acta Biomaterialia*, 2009, 5(8), 3098-3111.

[83] Morais, P.C., Silveira, L.B., Santos, J.G., Oliveira, A.C., Tronconi, A.L., Santos, R.L., Lima, E.C.D., Marchetti, J.M. and Tedesco, A.C. Use of the photoacoustic spectroscopy for surface characterization of nanometer-sized cobalt-ferrite particles. *IEEE Transactions on Magnetics*, 2005, 41(10), 3382-3384.

[84] Mungkalasiri, J., Bedel, L., Emieux, F., Dor, J., Renaud, F.N.R. and Maury, F. DLI-CVD of TiO_2-Cu antibacterial thin films: Growth and characterization. *Surface and Coatings Technology*, 2009, 204(6-7), 887-892.

[85] Muratore, C., Voevodin, A.A., Hu, J.J., Jones, J.G. and Zabinski, J.S. Growth and characterization of nanocomposite yttria-stabilized zirconia with Ag and Mo. *Surface and Coatings Technology*, 2005, 200(5-6), 1549-1554.

[86] Parlinska-Wojtan, M., Karimi, A., Coddet, O., Cselle, T. and Morstein, M. Characterization of thermally treated TiAlSiN coatings by TEM and nanoindentation. *Surface and Coatings Technology*, 2004, 188-189(1-3 SPEC.ISS.), 344-350.

[87] Perez, H., Nol, V., Cavaliere-Jaricot, S., Etcheberry, A. and Albouy, P.A. Nanocomposite Langmuir-Blodgett films based on crown derivatized platinum nanoparticles: Synthesis, characterization, and electrical properties. *Thin Solid Films*, 2008, 517(2), 755-763.
[88] Polychronopoulou, K., Neidhardt, J., Rebholz, C., Baker, M.A., O'Sullivan, M., Reiter, A.E., Gunnaes, A.E., Giannakopoulos, K. and Mitterer, C. Synthesis and characterization of Cr-B-N coatings deposited by reactive arc evaporation. *Journal of Materials Research*, 2008, 23(11), 3048-3055.
[89] Premkumar, P.A., Dasgupta, A., Kuppusami, P., Parameswaran, P., Mallika, C., Nagaraja, K.S. and Raghunathan, V.S. Synthesis and characterization of Ni and Ni/CrN nanocomposite coatings by plasma assisted metal-organic CVD. *Chemical Vapor Deposition*, 2006, 12(1), 39-45.
[90] Qi, Z.M., Honma, I., Ichihara, M. and Zhou, H. Layer-by-layer fabrication and characterization of gold-nanoparticle/ myoglobin nanocomposite films. *Advanced Functional Materials*, 2006, 16(3), 377-386.
[91] Qin, L., Cheng, H., Li, J.M. and Wang, Q.M. Characterization of polymer nanocomposite films using quartz thickness shear mode (TSM) acoustic wave sensor. *Sensors and Actuators, A: Physical*, 2007, 136(1), 111-117.
[92] Qufu, W., Yu, L., Ning, W. and Shanhu, H. Preparation and characterization of copper nanocomposite textiles. *Journal of Industrial Textiles,* 2008, 37(3), 275-283.
[93] Rahimi, A., Gharazi, S., Ershad-Langroudi, A. and Ghasemi, D. Synthesis and characterization of hydrophilic nanocomposite coating on glass substrate. *Journal of Applied Polymer Science*, 2006, 102(6), 5322-5329.
[94] Rahman, M.M., Kim, H.D. and Lee, W.K. Preparation and characterization of waterborne polyurethane/clay nanocomposite: Effect on water vapor permeability. *Journal of Applied Polymer Science*, 2008, 110(6), 3697-3705.
[95] Ramalingam, S., Muralidharan, V.S. and Subramania, A. Electrodeposition and characterization of Cu-TiO_2 nanocomposite coatings. *Journal of Solid State Electrochemistry*, 2009, 13(11), 1777-1783.
[96] Rao, Y. and Wong, C.P. Material Characterization of a High-Dielectric-Constant Polymer-Ceramic Composite for Embedded Capacitor for RF Applications. *Journal of Applied Polymer Science*, 2004, 92(4), 2228-2231.
[97] Ribeiro, E., Rebouta, L., Carvalho, S., Vaz, F., Fuentes, G.G., Rodriguez, R., Zazpe, M., Alves, E., Goudeau, P. and Rivi¨re, J.P. Characterization of hard DC-sputtered Si-based TiN coatings: The effect of composition and ion bombardment. *Surface and Coatings Technology*, 2004, 188-189(1-3 SPEC.ISS.), 351-357.
[98] Romero, J. and Lousa, A. OES time-resolved characterization of the deposition of multilayered Cr/CrN coatings. *Vacuum*, 2007, 81(11-12), 1421-1425.
[99] Rong, Y., Chen, H.Z., Wu, G. and Wang, M. Preparation and characterization of titanium dioxide nanoparticle/ polystyrene composites via radical polymerization. *Materials Chemistry and Physics*, 2005, 91(2-3), 370-374.
[100] Sadjadi, M.S. and Farhadyar, N. preparation and characterization of the hydrophilic nanocomposite coating based on epoxy resin and titanate on the glass substrate. *Journal of Nanoscience and Nanotechnology*, 2009, 9(2), 1172-1175.

[101] Sadjadi, M.S., Farhadyar, N. and Zare, K. Preparation and characterization of the transparent SiO_2 - Ag/PV P nanocomposite mirror film for the infrared region by sol-gel method. *Superlattices and Microstructures*, 2009, 46(3), 483-489.
[102] Sangermano, M., Amerio, E., Epicoco, P., Priola, A., Rizza, G. and Malucelli, G. Preparation and characterization of hybrid nanocomposite coatings by cationic UV-curing and the sol-gel process of a vinyl ether based system. *Macromolecular Materials and Engineering*, 2007, 292(5), 634-640.
[103] Sangermano, M., Amerio, E., Priola, A., Di Gianni, A. and Voit, B. Preparation and characterization of acrylic resin/titania hybrid nanocomposite coatings by photopolymerization and sol-gel process. *Journal of Applied Polymer Science*, 2006, 102(5), 4659-4664.
[104] Sangermano, M., Roppolo, I., Shan, G. and Andrews, M.P. Nanocomposite epoxy coatings containing rare earth ion-doped LaF3 nanoparticles: Film preparation and characterization. *Progress in Organic Coatings,* 2009, 65(4), 431-434.
[105] Schmid, A., Scherl, P., Armes, S.P., Leite, C.A.P. and Galembeck, F. Synthesis and characterization of film-forming colloidal nanocomposite particles prepared via surfactant-free aqueous emulsion copolymerization. *Macromolecules*, 2009, 42(11), 3721-3728.
[106] Sheen, Y.C., Lu, C.H., Huang, C.F., Kuo, S.W. and Chang, F.C. Synthesis and characterization of amorphous octakis-functionalized polyhedral oligomeric silsesquioxanes for polymer nanocomposites. *Polymer,* 2008, 49(18), 4017-4024.
[107] Silvain, J.F., Vincent, C., Heintz, J.M. and Chandra, N. Novel processing and characterization of Cu/CNF nanocomposite for high thermal conductivity applications. *Composites Science and Technology*, 2009, 69(14), 2474-2484.
[108] Sreedhar, B., Chattopadhyay, D.K. and Swapna, V. Thermal and surface characterization of polyurethane-urea clay nanocomposite coatings. *Journal of Applied Polymer Science*, 2006, 100(3), 2393-2401.
[109] Su, C., Wang, G. and Huang, F. Preparation and characterization of composites of polyaniline nanorods and multiwalled carbon nanotubes coated with polyaniline. *Journal of Applied Polymer Science*, 2007, 106(6), 4241-4247.
[110] Sukhodub, L. Materials and coatings based on biopolymer- apatite nanocomposites: Obtaining, structural characterization and in-vivo tests. *Materialwissenschaft und Werkstofftechnik*, 2009, 40(4), 318-325.
[111] Tanaka, T., Montanari, G.C. and Mulhaupt, R. Polymer nanocomposites as dielectrics and electrical insulation- perspectives for processing technologies, material characterization and future applications. *IEEE Transactions on Dielectrics and Electrical Insulation*, 2004, 11(5), 763-784.
[112] Tavman, I., en, V., zdemir, I., Turgut, A., Krupa, I., Omastova, M. and Novak, I. Preparation and characterization of highly electrically and thermally conductive polymeric nanocomposites. *Archives of Materials Science and Engineering,* 2009, 40(2), 84-88.
[113] Tedim, J., Gonalves, F., Pereira, M.F.R., Figueiredo, J.L., Moura, C., Freire, C. and Hillman, A.R. Preparation and characterization of poly[Ni(salen)(crown receptor)]/multi-walled carbon nanotube composite films. *Electrochimica Acta*, 2008, 53(23), 6722-6731.

[114] Thiemig, D. and Bund, A. Characterization of electrodeposited Ni-TiO2 nanocomposite coatings. *Surface and Coatings Technology*, 2008, 202(13), 2976-2984.

[115] Tseng, C.C., Hsieh, J.H., Wu, W., Chang, S.Y. and Chang, C.L. Surface and mechanical characterization of TaN-Ag nanocomposite thin films. *Thin Solid Films*, 2008, 516(16), 5424-5429.

[116] Tseng, H.H., Kumar, I.A., Weng, T.H., Lu, C.Y. and Wey, M.Y. Preparation and characterization of carbon molecular sieve membranes for gas separation-the effect of incorporated multi-wall carbon nanotubes. *Desalination*, 2009, 240(1-3), 40-45.

[117] Wang, X., Li, T., Lin, Q., Wang, D. and Zhao, T. Synthesis and characterization of ordered mesoporous carbon/silica hybrid thin films. *International Journal of Chemical Reactor Engineering*, 2008, 6.

[118] Wang, X.X., Li, T.H., Ji, Y.B. and Jin, W. Preparation and characterization of ordered mesoporous carbon membranes. Cailiao Gongcheng/*Journal of Materials Engineering*, 2008(11), 41-45.

[119] Wang, Y., Jiang, W., Zhang, X.F., Cheng, Z.P., An, C.W., Guo, X.D. and Li, F.S. Preparation and characterization of composite with micron Al coating in nanometer Fe particles. Gongneng Cailiao/*Journal of Functional Materials,* 2008, 39(11), 1900-1902.

[120] Wang, Z., Chen, X., Chen, M. and Wu, L. Facile fabrication method and characterization of hollow Ag/SiO_2 double-shelled spheres. *Langmuir,* 2009, 25(13), 7646-7651.

[121] Wei, Q., Yu, L., Hou, D. and Huang, F. Surface characterization and properties of functionalized nonwoven. *Journal of Applied Polymer Science,* 2008, 107(1), 132-137.

[122] Wei, Q., Yu, L., Mather, R.R. and Wang, X. Preparation and characterization of titanium dioxide nanocomposite fibers. *Journal of Materials Science*, 2007, 42(19), 8001-8005.

[123] Wu, X.L., Yue, T., Lu, R.R., Zhu, D.Z. and Zhu, Z.Y. Synthesis and characterization of the carbon nanotubes/magnesium oxysulfates nanocomposite by the hydrothermal reaction. *Chinese Journal of Inorganic Chemistry*, 2005, 21(4), 546-550.

[124] Xu, J., Yang, H., Fu, W., Fan, W., Zhu, Q., Li, M. and Zou, G. Synthesis and characterization of stainless steel/tin oxide: Bifunctional magnetic-optical nanocomposites. *Materials Science and Engineering B: Solid-State Materials for Advanced Technology*, 2007, 140(1-2), 132-136.

[125] Yang, B., Huang, Z.H., Liu, C.S., Zeng, Z.Y., Fan, X.J. and Fu, D.J. Characterization and properties of Ti-containing amorphous carbon nanocomposite coatings prepared by middle frequency magnetron sputtering. *Surface and Coatings Technology*, 2006, 200(20-21), 5812-5818.

[126] Yeh, J.M., Kuo, T.H., Huang, H.J., Chang, K.C., Chang, M.Y. and Yang, J.C. Preparation and characterization of poly(o-methoxyaniline)/Na^+-MMT clay nanocomposite via emulsion polymerization: Electrochemical studies of corrosion protection. *European Polymer Journal*, 2007, 43(5), 1624-1634.

[127] Yeh, J.M., Yao, C.T., Hsieh, C.F., Lin, L.H., Chen, P.L., Wu, J.C., Yang, H.C. and Wu, C.P. Preparation, characterization and electrochemical corrosion studies on environmentally friendly waterborne polyurethane/Na^+-MMT clay nanocomposite coatings. *European Polymer Journal*, 2008, 44(10), 3046-3056.

[128] Yuan, G.L., Li, W.M., Yin, S., Zou, F., Long, K.C. and Yang, Z.F. Nanocomposites of urethane and montmorillonite clay in emulsion: In situ preparation and characterization. *Journal of Applied Polymer Science,* 2009, 114(3), 1964-1969.

[129] Zhang, M., Shi, L. and O'Connor, C.J. Preparation and characterization of silica-coated indium oxide nanoparticles. *Journal of Nanoscience and Nanotechnology*, 2008, 8(11), 5720-5724.

[130] Zhao, D., Zhou, J. and Liu, N. Characterization of the structure and catalytic activity of copper modified palygorskite/TiO_2 (Cu^{2+}-PG/TiO_2) catalysts. *Materials Science and Engineering* A, 2006, 431(1-2), 256-262.

[131] Zheng, H.y., An, M.z. and Lu, J.f. Surface characterization of the Zn-Ni-Al_2O_3 nanocomposite coating fabricated under ultrasound condition. *Applied Surface Science*, 2008, 254(6), 1644-1650.

[132] Zhu, Y.C. and Jiang, J.S. Synthesis and characterization of Fe_2O_3-ATO nanocomposite particles. *Surface Review and Letters*, 2008, 15(5), 545-550.

Chapter 4

CLASSIFICATION AND STUDY OF NANOCOMPOSITE COATINGS

ABSTRACT

Some classifications and studies of nanocomposite coatings were discussed in this chapter. It was focused on an example of nanocomposites which is Ni/SiC nanocomposite coating. Different reinforcing particles such as oxides, carbides, nitrides, sulfides, polymers, graphite and nano-fibers and carbon nano-tubes, metallic particles and diamond - depending on their particular characteristics - are added to metallic matrix. Regarding their expected properties and applications, these particles may have dimensions about several nano-meters to several micrometers. These reinforcing particles may have one or more of mentioned characteristics. However, achieving desirable properties through one particle type may result in reducing another capability of the coating. For instance, adding Al_2O_3 and TiO_2 particles for enhancing nickel coatings' hardness and abrasion resistance leads to decrease of wear strength in a chlorine content environment, since nickel particle interface is a suitable path for diffusion of invader ions, such as chlorine ions, into the coating.

4.1. INTRODUCTION

For its variety of resulted properties, efficient controllability, low costs, and low working temperature, electrochemical deposition has been regarded as one of most applicable and useful methods for variant usages. Along expanding and improving coatings features - considering their usages - there is a continuous effort for producing new coatings with optimum required characteristics. Composite coating, and particularly deposition through simultaneous electro-deposition of matrix and reinforcing phase, is among innovations for producing coatings with higher efficiency and characteristics. At first years of electro-coating, yielded coatings were highly rough and irregular due to entrapment of exterior constituents. These constituents were presented as insoluble particles in electrolyte and readily introduced into coatings. The particles real source was impurities resulted from chemical composition used in cathodes and anodes.

Nowadays this problem is solved with access to chemicals and anodes of high purity. However, electrolyte capability for simultaneous deposition of exterior particles is not neglected. Currently, this capability is used for producing composite coatings with desirable pure particles at controlled condition, for achieving given physical and mechanical properties. Although, more than 30 years have passed from the first introducing of electrochemical composites, studies in this area is continuously followed by many researchers. Combining metallic matrix properties such as strength, efficient coherence, capability of reshaping, and etc. with particles variant reinforcing properties leads to yield coatings with desirable characteristics.

4.2. Composite Coatings

The objectives for producing composite coatings include enhancing hardness, increasing abrasion resistance against electrolyte wear and high temperature oxidation and wear, self-lubrication, producing coatings with special composition after simultaneous composite deposition, and finally medical applications. Different reinforcing particles such as oxides, carbides, nitrides, sulfides, polymers, graphite and nano-fibers and carbon nano-tubes, metallic particles and diamond - depending on their particular characteristics - are added to metallic matrix. Regarding their expected properties and applications, these particles may have dimensions about several nano-meters (such as Al_2O_3 with average size of less than 14 nm in nickel matrix) to several micrometers (such as diamonds with average size of 150 microns particles in same matrix). These reinforcing particles may have one or more of mentioned characteristics. However, achieving desirable properties through one particle type may result in reducing another capability of the coating. For instance, adding Al_2O_3 and TiO_2 particles for enhancing nickel coatings' hardness and abrasion resistance leads to decrease of wear strength in a chlorine content environment, since nickel particle interface is a suitable path for diffusion of invader ions, such as chlorine ions, into the coating.

4.3. Composite Coatings Application

For their high degree of hardness, ceramic particles are generally added to coating to increase its hardness and abrasion resistance, besides improving its mechanical characteristics. It is worthy to say that these particles presence usually conducts in improving wear strength. First group, which are of a high interest, is oxide particles. Until now different oxide particles are successfully subjected to simultaneous electrolyte deposition. Al_2O_3 is the most considered member of this group. Adding Al_2O_3 to nickel matrix leads to an enhancement in hardness, abrasion resistance and friction ratio of the coating. Studies about these particles two-staged deposition in a P-Ni alloy matrix include:

1. Electrophoretic deposition of Al_2O_3 particles
2. Electrolytic deposition of nickel ions and composite development.

These two steps obviously show that this process makes better tribological characteristics, since it creates stronger bonds between Al_2O_3 particles and metallic matrix. Al_2O_3 particles are successfully deposited in Ni-Fe alloy matrix. Adding Fe ions Al_2O_3

deposition rate will noticeably increase, for an increase in surface charge through Fe ions adsorption.

Other oxide particles such as TiO_2, ZrO_2, ZnO, and SiO_2 are also used to improve coatings surface characteristics. ZnO particles, as well as increasing Ni-P coatings hardness up to 400 Hv, noticeably decrease E_{ocp} and anodic polarization, as well as offering favorable wear strength. TiO_2 particles have also an important effect on hardness and abrasion resistance. These particles presence causes a decrease in crystallization temperature up to 100^0C, which is all in all resulted in rising of network defects at presence of second phase particles. On the other hand, these particles serve as a lock to stop matrix grains during recrystallization. Hence, in addition to creating a possibility for an increase in stress capacity and ductility of the coatings, it will be feasible to obtain fine-grained and coarse-grained coatings. ZrO_2 and SiO_2, also, make a hard fine-grained coating which has high wear strength.

Carbides are another important and widely used group of ceramics. Besides SiC, carbides composite deposition with WC and B_4C hard particles are yielded. For its high hardness and abrasion resistance capability, WC particles are used in nickel coatings. B_4C particles deposition is different from the other ceramics. The most noteworthy difference between B_4C with the other ceramics is its electrical conductivity and hydrophobic properties. For this reason B_4C deposition from bathes with low concentration is possible. Also, the particles within the coating will be accommodated with higher stability and will offer better surface characteristics.

Nitrides and Fluorides are also applied as reinforcing particles within nickel matrix. Silicon Nitrides, Offering an efficient hardness of 500 Hv, and CaF_2 having a negligible friction ratio, provide desirable surface characteristics.

For its great hardness, providing composites with diamond as reinforcing phase is absolutely desirable. Hence for abrasion resistant consumptions nickel composites are made with diamond particles. Regarding this issue, depositing very hard and abrasive particles of diamond and cBN within nickel matrix on steel gears of the grinding machine, it is possible to make a grinding machine which is capable to abrade very hard materials- impossible with ceramic or steel ones.

Having self-lubricating property, low friction rate, extremely low cohesion, and lacking in deposition adsorption, Poly Tetra Flouro Ethylene (PTFE) particles produce very efficient coatings for abrasive usages and boilers' containers coating. Simultaneous deposition of PTFE particles considerably decreases the chance and number of developed cracks in nickel coating. Poly Ethylene (PE) is another polymeric particle which is successfully deposited as reinforcing phase. These particles enhance nickel coatings wear resistance and build up it up to 600 Hv.

For their high tensional strength and desirable mechanical properties with efficient self-lubrication, carbon nano-tubes are one of the other proposed candidates for simultaneous deposition with metallic phase. Nickel composite and Carbon nano-tube (CNT) composites are of less friction ratio compared with nickel coating and Ni-SiC composite. Hardness of this composite, being up to 850 Hv, is also highly noticeable. Thus, self-lubrication and low friction ratio of this composite demonstrates a considerable wear resistance, as its wear resistance is even higher than that of Ni-SiC and Ni-graphite composites. Graphite is another self-lubricant particle which is widely studied and used in composite coatings. Graphite composite within metallic matrix offers desirable wear property besides low friction ratio.

Simultaneous deposition of metallic particles within metallic matrix is of high application. One of key objectives is to produce a coating with a new inter-metallic phase. As an example for this, one can name aluminum particles simultaneous deposition within nickel matrix for producing inter-metallic Ni_3Al coating. This coating is applied in high temperature operations since it can save permanent stability, noticeable oxidation strength, and a high stability to weight ratio.

After deposition, in the second phase, thermal operation is executed at 500 to 800 °C, leading to development of Ni_3Al inter-metallic phase. However, here metal-metal composite coating is focused. The following examples may help to understand the issue:

- simultaneous deposition of Ag-Sn alloy with Ag particles for improving surface solder quality in final solder operation in absence of Pb
- Simultaneous deposition of Al and Ti particles within Ni matrix which ends up in producing composites with one or two reinforcing phase
- Niobium (Nb) particles which enhance wear strength of the coating, as well as its hardness, due to its high strength. Also Nb particles enhance the coating strength through changing pillar structure of Ni coating into nano-structure (decreasing corrosion current density of Ni coating from 127 to 30 $\mu A/cm^2$).

Another interesting process is simultaneous inter-metallic deposition with two reinforcing particles. As examples one can refer to simultaneous deposition of Al particles with Se or Y (Yttrium) within Ni matrix, where after thermal operation, finally, there develop a coating of oxide particles within Ni_3Al inter-metallic phase which is of a high strength against wear and oxidation. Also, Simultaneous deposition of B amorphous particles within Ni matrix is performed for enhancing coatings' hardness and strength against abrasion.

Expanding the range of the studies on composite coating, this method was applied for implants coating. Combining bio-compatibility properties of hydroxiapatite (Hap) with Ni strength and ductility within a composite made of this two on stainless steel is another new result of this study. In this way Hap's brittleness problem is solved through changing its combination and structure and via its high temperature decomposition. As another instance one can name these composite coatings' medical usages in producing sensors which react to presence of Acetylcholine. Simultaneous deposition of NiO within Ni matrix results in producing such sensors. It is worthy to mention that an unusual decrease in human body's Acetylcholine can cause diseases such as Alzheimer's and Parkinson. Thus, measuring the amount of this substance is of a great importance.

In nuclear plants cooler tubes surface is in touch on side with plasma phase and in the other side with heat absorbing system. Hence, they must be capable of bearing intense stresses caused by high thermal gradients, as well efficiently fast transfer of the heat. Also, these tubes must maintain their mechanical characteristics in these conditions. Such characteristics can be obtained through combining coppers' high temperature heat conductivity with SiC's strength and stability. For this reason in such structures it is preferable to use coatings composed of copper composite with SiC's reinforcing fibers. Besides composite coating, simultaneous composite deposition of metal-particle can also be applied for other aims such as producing composite powders. Successful producing of copper composite powder, including carbon nano-fibers for secondary usages in powder metallurgy, is categorized in this group.

4.4. Ni-SiC Composite Coating

Hard particles of SiC are one of most commonly used reinforcing particles in composite coatings. Here, desirable properties of adding these composites into a nickel matrix, as well as its low costs, are of great interest. One may say these composites are mostly considered composites. Some of these composites characteristics will be discussed as follows:

4.4.1. Hardness

Since SiC particles are very hard ones, they are added into Ni metallic matrix as reinforcing phase, to enhance coatings' hardness and strength against abrasion. All researchers agree about the relationship between hardness and SiC content of a coating.

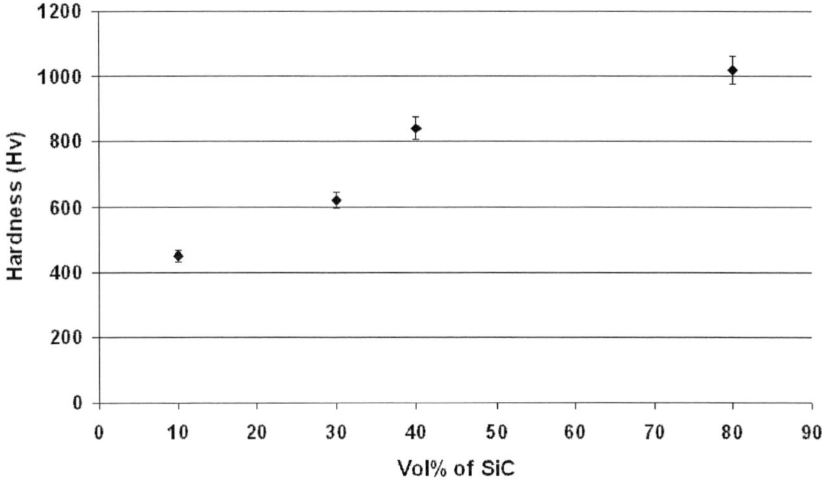

Figure 4.1. the effect of SiC volumetric percent on coatings' hardness.

This has been reflected in many articles. As it can be seen in figure 4.1, the process of hardness increase with amount of SiC continues till reaching to optimum hardness of 1050 Hv, where SiC content is 80% of coating volume.

The reason for this increase can be summarized in cooperation of three different processes:

A. Coatings stimulation to develop in hard crystallography directions: crystallographic studies on pure Ni crystalline deposition reveals that these metal deposit growth is visible in XRD photos at preferable orientation of [100] and [200], while through SiC micro-particles deposition and under influence of these particles crystallography orientation gradually changes. So, in XRD diffractions of the coating [100] lines gradually get dim and weak but [311] and [111] lines get darker and stronger. Therefore, there will be micro-structures with orientation which a combination of 100 and 211 directions. As far as it is known, the microstructure resulted from [100] orientation will create deformable sediment with lowest hardness and internal stresses. On the other hand, the micro-structures with [100] and [211] orientation will create rigid sediment with low deformability and high internal stresses. This process results into developing hard sediment. Structure modification is a process which is common in composites with micro size particles rather than nano ones. This is due to incapability of very tiny nano-meter sized SiC particles in disturbing order of nickel coating network.

B. Grain size decrease: an increase in hardness because of a decrease in grain size happens according to Hall-Petch mechanism. Based on Hall-Petch equation hardness has a direct relationship with -0.5 exponent of grain size. Indeed, one can introduce an equation for hardness, same as strength, as follows:

$$H_1 = kd^{-1/2} + H$$

Considering the equation above it seems that Ni grain size decrease can highly affect composite coating hardness. For this reason, in composites with nano-crystalline matrix, SiC content changes and presence or absence of SiC - particularly at scales larger than nano - has a little or no effect on coating hardness. On the other hand, as we know, adding particles into the network results in a noticeable increase of metallic matrix fine grains. For instance, a study shows that adding SiC to coating reduces the size of nickel particle from 0.23 μm to 78 nm, without changing other parameters.

C. Diffusional hardness due to presence of hard particles on the surface: According Orwan mechanism if there are hard particles, they will serve as a barrier against dislocation movements during plastic deformation. Inserted stress on each dislocation results in dislocations bending around particles and, finally, passing the dislocation with leaving a dislocation loop around the particle. The loop will apply a mutual stress to dislocation sources. Then, this stress must be override for occurrence such a slide. For the sake of this mechanism an in presence of hard dispersive particles the under stress structure will quickly harden. Based on Orwan's theory, required stress for passing dislocation from the particle is inversely related with particles distance.

$$\tau_0 = \frac{Gb}{\lambda}$$

It is obvious that with an increase in hard particles numbers, particles inter-space decreases. Then hardness would increase according to this mechanism. Regarding the abovementioned points it sounds there are some contradiction in some cases, particularly when it comes to compare composites with different particle dimensions. For example experiments reveal that a little amount of submicron particles, sometimes, has a greater effect than that of larger than micron particles amount. In research, though SiC weight percentage for nano-size particles was ranged 2.5 to 6.46% was yielded versus 0.5 to 1.7 weight percentage for nano particles, but numerical density of these particles within the coating is very different. It varies form 2.23×10^{15} to 11.18×10^{15} particles per cm^3. On the other hand the density for micron size particles is varied between 12.12×10^{10} and 30.80×10^{10} particle per cm^3. Needless to say this egregious difference in number of particles leads to highly reduction of particles' inter-space, increase of dislocation locks, and consequently Orwan mechanism domination. Also, this difference in number results in greater hardening of composites with nano particles, in fairly smaller percentages of composites with micro particles. On the other hand, micro particles with larger dimensions disturbs nickel coating structure and stimulate orientation in [211] direction with greater ease, during accommodation within the network. However, in composite coating with nano-particles the former orientation of nickel coating ([100]) will be dominant.

However, adding micro-particles or submicron particles to nano-crystalline network has an considerable effect on hardness. It must be noted that in such networks particles size is larger than that of grains. Then, each particles will surrounded with a high number of nano-crystals and model of dislocation accumulation in backside of grains border (Hall-Petch model) will be most efficient one for enhancing the hardness in such structures. Here the other noteworthy point during study of coating hardening mechanism with micro and nano particles of SiC is that micro-particles deposit in inter-crystalline form at edges and corners of nickel crystals, while SiC nano-particles have intra-crystalline subsidence behavior can be deposited into Ni grains, as well as grains' borders. Therefore, it can be acclaim that Orwan hardening mechanism or diffusional hardness will be shown where nano size particles are used in composite. Nevertheless, the mechanism of hardening due to orientation change is more frequent in micro-particles and Hall-Petch mechanism will exists in variant grain sizes.

4.4.2. Wear Resistance

As previously discussed, the most important aim of adding SiC particles into Ni matrix is enhancing its wear strengths. One of most basic reasons of this enhancement is an increase in coating hardness, which previously discussed in detail. Results of variant researches also prove this parallel increase of wear strength with hardness. For instance one can refer to results, which found highest degree of wear strength is corresponded with highest amount of SiC and greatest hardness. A comparison between the curves offered in figure 4.2 is an evidence for this claim.

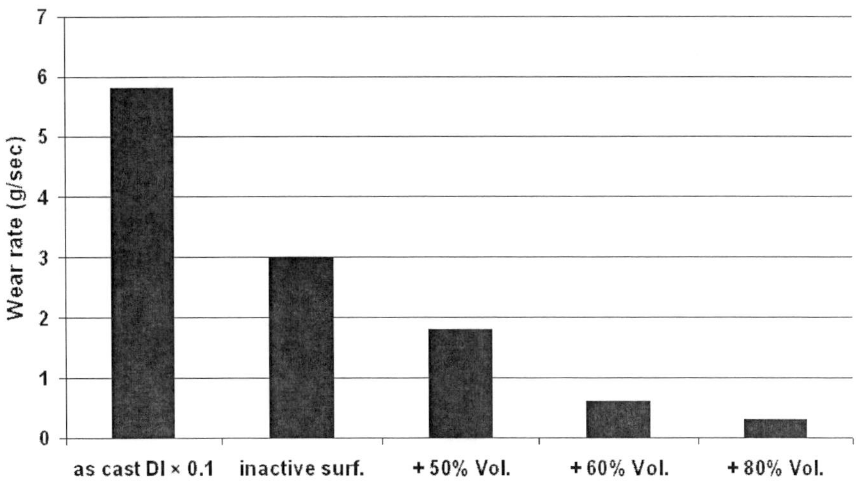

Figure 4.2. wear changes rate with coatings' SiC content.

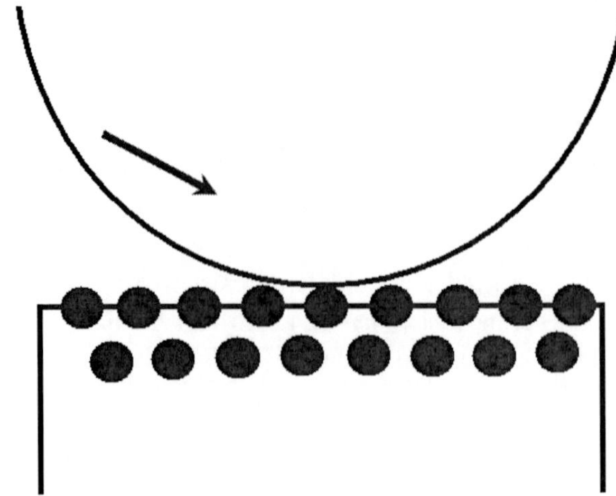

Figure 4.3. Schematic figure of SiC hard particles effect on wear strength.

Numerical density of SiC particles has another key role as well as hardness enhancement. As numerical density increases inter-particle spaces decrease. In low densities abrasive substance readily pulls apart Ni coating, due to these particles large distances. As well as Ni some particles also dislodge from the coating and only few particles are left uncovered on its surface.

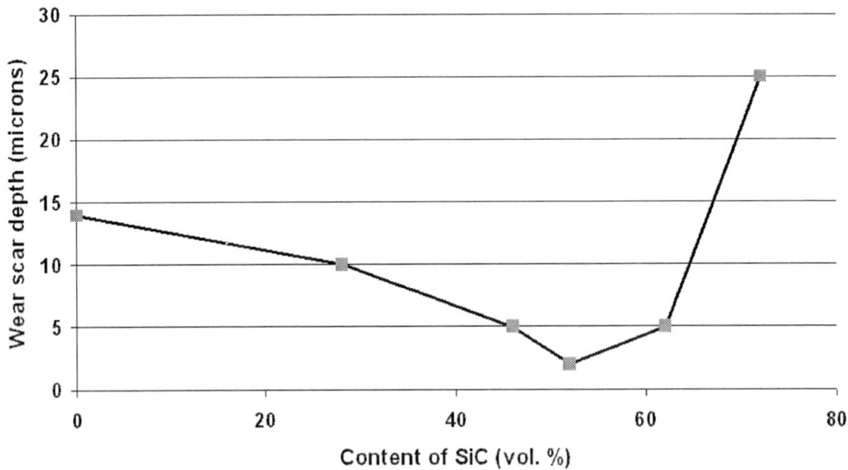

Figure 4.4. SiC content effect on coating's wear strength.

These particles will be dislodged with a great ease in subsequent cycles. Further, once particles numerical density within the coating is low, particles close distances will lead to an efficient coverage of the particles and prevents Ni abrasion and dislodging from surface, which in turn cause to preservation of particles strength in presence of SiC particles and an acceptable strength for the cover (figure 4.3). Hence, few amounts of submicron particles have a similar effect with high number of micro-particles on wear strength.

Another important effective factor on composite coatings wear strength is particles strength and cohesion. However, in Ni-SiC coatings the particle with strong bounds with Ni will remain on surface. This is assumed one of reasons for the wide application of SiC in enhancing Ni wear strength. As it previously mentioned, an increase in SiC content of the coating causes an increase in wear strength. Researches show that this increase continues until reaching an optimum volumetric percentage of SiC and after that it will decline. This may be due to incapability of the small number of remained nickel particles on the surface for preserving SiC hard particles into the coating. Some researchers recorded the number of 52 volumetric percentages for SiC optimum amount (figure 4.4). However, some other researchers have even recorded this strength increase rate up to 80 volumetric percentages.

4.4.3. Friction Coefficient

Coating's friction coefficient is suggested as an important factor in its surface behavior. There are so many records about friction coefficient behavior of composite coating, which sometimes are in contradiction with each other. Some researchers have recorded a friction coefficient for composite higher than that of Ni coating. There are other records which imply friction coefficient is equal for both composite and pure coatings. Even, some opponents believe that composite friction coefficient is less than that of pure Ni coating. However, all researchers agree in these two following points:

1. A coating made of nano-particles have smaller friction coefficient compared to micro-particles.
2. Composite coating has a more stable friction coefficient in comparison with pure Ni coating.

Different governing condition on experiments, including dimensions, particles different ratios, different amount of applied loads during the tests, different investigation times, and finally different selected environments can be assumed the source of this differences.

In low rates of SiC, friction coefficient will be highly changing during abrasion and loading time. This phenomenon, which occurs in particles large intervals, is due to changes of coating surface and can be seen at 8 volumetric percentage of SiC. After that the changes decreases and the curve has a stable form. Pursuing the test and in higher contents of SiC, the curves' changes with Friction coefficient decrease and it would be in a stable shape. Compared to composite coating, Ni coating friction coefficient is highly depended on time and abrasion rate and there would be great changes in it. However, it is expected to have a stable behavior in composite state. The most important factors for this difference are surface morphology and cohesive abrasion's key effect on Ni coatings.

Also, friction coefficient dependency of composite on inserted load is less than pure Ni coating's. At low rate of applied loads, friction coefficient of composite coating is lower than that of pure Ni coatings. Pure Ni coating is of a rough surface with coarse grains but composite coating is somehow compacted and fine grained. For this reason, as it goes on at first stage of loading friction coefficient of pure Ni coating quickly declines, since with coating surface deformation and more compaction through applied load the surface will be of a better quality. However, at higher applied loads, although Ni surface's improvement is performed with higher speed, but SiC particles are engaged with abrasive particles. Hence, composite's friction coefficient increases and overrides that of pure Ni coating.

The higher required time for friction coefficient for being stable, particularly at composites, may be due to abrasive's surface wear and remaining FeO on coating-abrasive interface. The oxidized iron on the interface is deposited as a layer on the surface and contributes to surface to be protected against abrasion - through its self-lubricating feature.

4.4.4. Corrosion Resistance

Increasing SiC particles content enhance corrosion resistance of the coating, as well as an increase in its abrasive characteristics. Results of salt spray for Ni and composite coatings are introduced in figure 4.5. It is obviously seen that over the time composite coating is of greater resistance. In another test pure Ni and composite coatings are developed at a cylindrical sample and the sample is orbited in a 0.5 molar sodium sulphate solution with different speeds. At first E_{corr} = -260 mV was for pure Ni coating, while for composite coating we have E_{corr} = -198 mV. As rotation speed goes up, corrosion potential of composite sample moves to higher positive values. On the other hand Ni sample demonstrates higher negative values. Table 4.1 shows different measured parameter values at rotation speed of 150 rpm. All values imply a considerable improvement at wear strength of the coating while adding SiC particles.

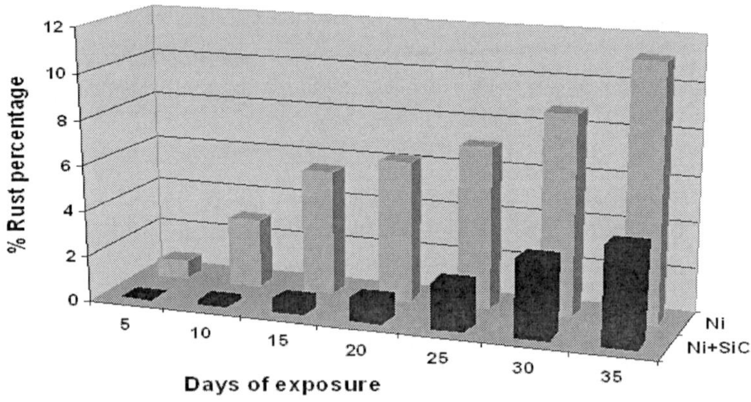

Figure 4.5. Formed corrosion products on composite coatings under salt spray test.

Table 4.1. Potential, corrosion current, cathodic and anodic tafel slopes for pure Ni coatings and Ni-SiC

Type of coating	E_{corr} (mV (Ag/AgCl))	i_{corr} (μA.cm^{-2})	β_a (mV per decade)	β_c (mV per decade)
Ni + SiC	-212.5	1.9	217	236
Pure Ni	-281.6	4.1	76	374

When SiC particles are embedded within Ni coating its corrosion resistance enhances through a change in morphology and reduction of Ni grains size. On the other hand, SiC particles within the coating react with corrosive environment of passive layer and develop SiO and Si (OH) protector layer. Developing such a protector layer the surface will not be corroded anymore. As well as abovementioned points, SiC particles protect the surface through covering a part of surface and preventing the metal contact with environment.

4.4.5. Strength against Wear-Corrosion

Studies results suggest that corrosion potential of composite coating under applying the abrasion is about Ni's passivation potential. However, only after an increase in applied load during abrasion there would be a small decrease of the potential. This implies that corrosion is not the cause of change in composite coating's protecting condition. On the other hand in pure Ni coating through load apply the passive layer on the coating rapidly disappears. Then, Ni's porous coating made it possible for aggressive ions to enter to the coating and reach to sub-layer. Nonetheless, SiC hard particles in the composite coating prevent passive layer from damages. The passive layer in Ni coating includes NiO and $Ni(OH)_2$. SiC nano-particles are larger than this layer and stop abrasives to touch passive layer.

Wear-corrosion strength of composites and pure Ni, under different applied loads, are introduced in a bar chart at figure 4.6. A comparison between the charts shows that wear-corrosion strength of composites is higher than that of pure Ni.

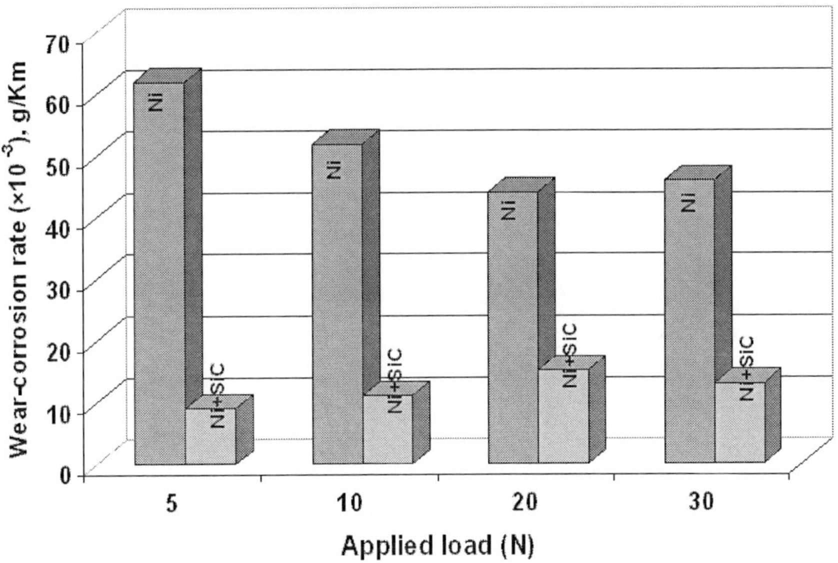

Figure 4.6. Intensity of wear-corrosion of pure and composite Ni coating under different loads at 0.5 molar sodium sulphate solution.

4.4.6. Mechanical Properties

At presence of SiC particles yield stress and ultimate strength of the Ni coating has a significant increase. For example, adding SiC submicron particles into nano-crystalline matrix of the Ni causes coatings yield strength to increase from 630-851MPa to 730-1050MPa for different nano-crystalline samples. Also, their ultimate tensional strength increases from 641-1056MPa to 995-1431MPa. In normal Ni coatings with larger grain sizes these increase has a more intense rate since yield stress of an anile pure Ni coating is about 180MPa while this is about 600 MPa for composite coating (figure 4.7). Size reduced SiC particles also leads to an increase in tensional strength.

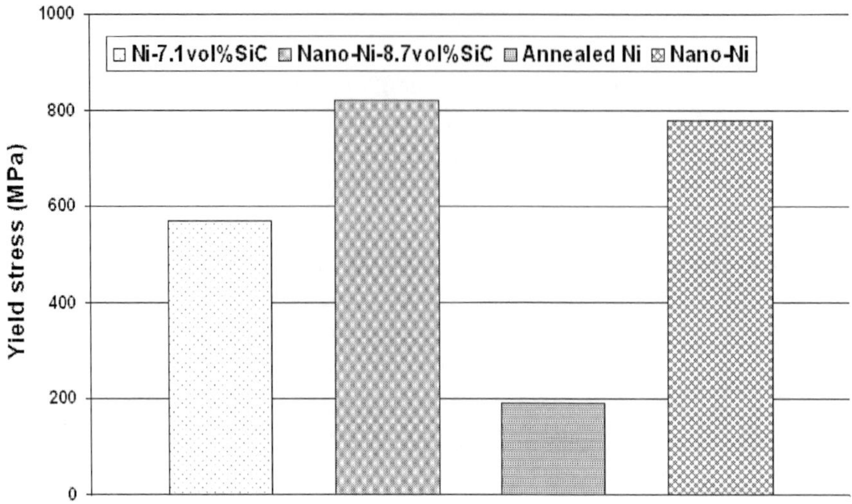

Figure 4.7. Comparison of yield stress for different coatings.

Yield stress and tensional strength's intense increase in common composites, is for these coatings strength increase, due to dispersed SiC particles in it. In other words, in these coatings SiC particles serve as obstacles against dislocation movements to prevent plastic deformation. On the other hand, in nano-crystalline composites, plastic deformation is limited by Hall-Petch mechanism. For this reason in these composites SiC increases does not have as great effect as that of on micro-crystalline coating. Here it must be noted that coatings strength decreases with SiC content increase. SiC particles agglomeration in the coating may be the answer of the question. Then, there will remind some pores between SiC particles which lead to weakening coatings strength.

On can assume three kinds of cracks in composite coating:

1. SiC particles cracking
2. Particles lodging from metallic matrix
3. Crack creating in matrix

Increasing applied stress the metallic matrix bears a deformation while SiC particles remain intact. Hence, for reducing internal stress particles and matrix segregation is necessary. Once, particle-matrix bound is not efficiently strong, all applied stresses convey to the particle. Hence, the fracture simply occurs whenever applied stress reaches to SiC fracture stress. The coating's thermal operation may have an important effect on its tensional strength. As it can be seen in figure 4.8, thermal operation of the coating at temperature of 300°C for 24 hours enhances sample's tensional strength up to 250MPa. Reinforcing particle-matrix bound through eliminating pore spaces at their interface and removing networks defects technically causes two cracking possibilities out of three be removed.

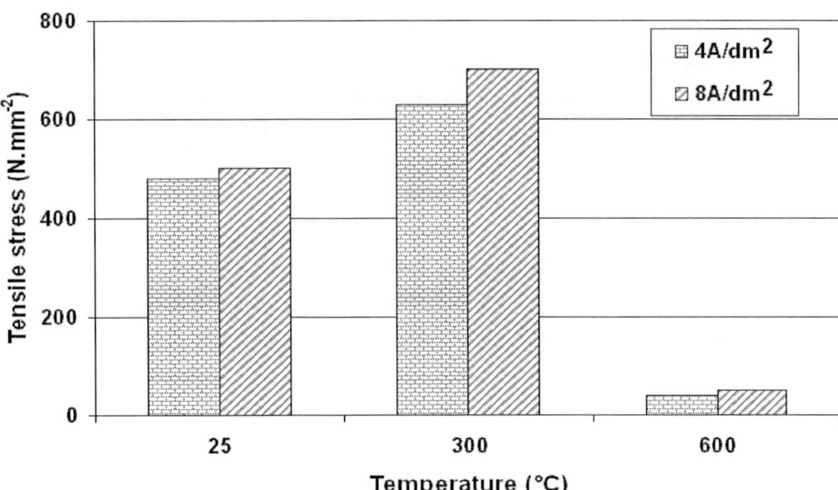

Figure 4.8. Tensional strength changes with thermal operational temperature of produced composite coating under $4A/dm^2$ and $8A/dm^2$.

Nevertheless, through performing this operation at 600°C the strength dramatically drops. Increasing the temperature up to 600°C Ni penetrates into SiC network, eliminates bounds between Si and C, and develops graphite and silicate phases (according to equation below), and remaining many shrinkage cavities causes strength declination:

$$3Ni + SiC \rightarrow Ni_3Si + C$$

Length change before the fracture for composite coatings is more than that of pure nickel coating. SiC content increase, due to these particles agglomeration, reduces final length changes. Ni-SiC composite coating can suffer a length change at environment temperature of 20 to 30 percent and strain rate of $8.33 \times 10^{-4}\ s^{-1}$. This length change at higher temperatures can be exhibited as super-plastic behaviors. The maximum of 571% elongation has been recorded at 703°K.

4.5. Production Methods

Different methods of composite coating are based on particles suspension within the electrolyte during deposition. Therefore, electrolyte turbulence is an inseparable part of these coatings. With this approach, technically these coatings production methods are categorized into three methods, including: conventional method, deposition method, and rotating electrode method.

4.5.1. Conventional Electro Co-Deposition (CECD)

By this method insoluble particles are suspended within conventional electrolyte by its agitation. There is a set-up of an electrical current between vertical and parallel anode and cathode. Composite coating is conventionally deposited on cathode, same as applied process in pure metal coating. The process is performed at high concentrations of the bath (100 to 150 g/lit). During the process the electrolyte is always in tribulation, which contributes to particles permanent suspension. This tribulation at bath is applied through different methods such as agitation using the air, magnetic agitation, vibratory agitation, mechanical agitation, or solution pumping. Here, magnetic agitation and agitation through electrolyte pumping have most practical method in CECD method.

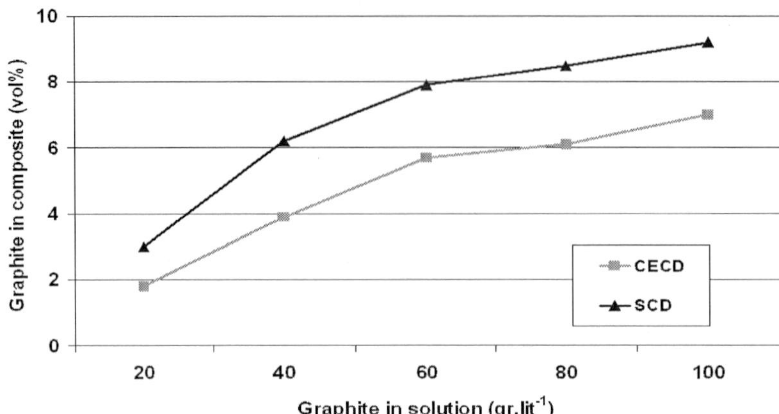

Figure 4.9. Comparison of copper coatings' graphite content for both SCD and CECD methods.

4.5.2. Sediment Co-Deposition (SCD)

At SCD method the anode is installed at top of the bath and cathode is placed horizontally under the anode. During deposition process electrolyte is intensely agitated at definite times. After complete suspension electrolyte particles are left by themselves. While particles are settled down, the particles will be placed at horizontal cathode through gravitational force. For this reason there will be developed a high number of particles with weak and strong bounds on the cathode, so the particles will be stock and entrapped within metallic developing matrix.

Through this process, as well as using low concentrated bathes, it is possible to obtain composites with very high volumetric concentrations, higher uniformity, and less porosity compared with CECD method. Figure 4.9 shows graphite volumetric percent within Cu coating for same concentrations, through both CECD and SCD method. It is revealed that the volumetric percent obtained from SCD method is more than that of CECD method. However, particles produced through SCD method are of more roughness than those obtained from CECD method.

4.5.3. Rotating Electrode Method: Cathode Rotating (CR) and Anode Circumference Rotating (ACR)

Cathode rotating is performed in both CR and ACR methods. Indeed rotating electrode method is one of CECD's subgroups since in this method, similar to CECD, vertical anode and cathode are used. Cathode rotation around itself causes a more unified thickness and better dispersion of particles within the composite, as well as electrolyte agitation. Cathode rotating around itself and anode in CR and ACR methods leads to electrode agitation. This makes it possible to particles to be buoyant and more mobile. Then, it will be more chance of developing ionic cloud on suspended particles. On the other hand, cathode movement causes a decrease in dual layer between cathode and anode, which in turn will move system's limit current. For two abovementioned effects composite deposition process is facilitated and the ratio of particles within the coating. As a result in same working conditions coating's hardness, will override that of a coating developed through CECD method.

4.6. Plating Conditions

In this section, after introducing consumed electrodes in Ni-SiC composite coating, different factors effect on production and their effect on the coating will discuss.

4.6.1. Electrolytes

For electro-deposition these following methods are commonly used:

1. Watts bath 2. Sulphamate bath 3. Sulphate bath 4. Chloride bath 5. Flouro-borate bath

Among these baths, Watts bath is the most commonly used. In this bath $NiSO_4$ is used in noncomplex form as a source for supplying needed nickel ions for the bath. Other components of the bath are:

- $NiCl_2$: for chloride ions compatibility in breaking barrier layer on Ni anode and improving its solubility and increasing density of limit currents through breaking intermediate layer.
- H_3BO_3: used for maintaining bath pH around 4 - and consequently reducing affinity to hydrolyze - and decreasing the rate of basic salts development.

Sulphamate bath and the other applied bathes have a composition somehow similar to Watts bath composition, except their Ni ions source is different. For instance, in sulphamate bath Ni $(NH_2SO_3)_2$ is the source to supply Ni ions. Chemical components and additives concentration and the other factors of Ni deposition bath are defined regarding considered characteristics. For instance, the following bath is recommended for electroforming:

$NiSO_4$ (gr/lit): 195 to 240

$NiCl_2$ (gr/lit): 45 to 75

H_3BO_3 (gr/lit): 22 to 30

Temperature: (°C) 40 to 45
Current density (A/dm^2): 0.8 to 4.4
pH: 4.8 to 5.4

For composite deposition Watts and Sulphamate baths, with changing their different parameters, are commonly used. Watts bath is proposed as most practical one in Ni composite coating. The applied Watt baths have the following composition:

$NiSO_4$ (gr/lit): 240 to 300

$NiCl_2$ (gr/lit): 35 to 45

H_3BO_3 (gr/lit): 30 to 45

However, sometimes different concentrations are also used, e.g. a bath with 60gr/lit concentration. Sulphamate applied baths for composite coating is more extensive than Watts baths. In these baths, Acid boric, and $NiCl_2$ are presented, as well as nickel sulphamate.

Nickel sulphamate (gr/lit): 70 to 400

$NiCl_2$ (gr/lit): 3 to 15

H_3BO_3 (gr/lit): 30 to 40

Controlling pH during plating acidic or basic additives are used, e.g. for reducing the pH, HCl, H_2SO_4, H_3BO_3 and a 1:7 combination of HCl and H_2SO_4. Also increasing the pH one can use $NiCO_3$ or NaOH.

As well as mentioned combinations, one can also add SiC particles to the bath. Other additives such as surfactants are also applied.

Nevertheless, for Ni-SiC composite deposition nonaqueous solutions are used as well as aqueous ones. Since particles hydrophobic and hydrophilic behaviors have effect on their deposition behaviors, and regarding that SiC particles are hydrophilic, the aqueous layer placed on the surface will serve as an obstacle against simultaneous deposition of these particles within the coating.

So, it seems that using aqueous baths will have an effect on simultaneous deposition. Hence, a nonaqueous bath with ethanol solvent with purity of 99.95% and water content of 0.4% was used. The applied solution's combination was as follows:

200 g/l $NiCl_2(H_2O)_6$, 0.11 mol in one liter of HCl, and 20gr/l SiC particles with average diameter of 1 micron. For ethanol vaporizing each 5 minutes some ethanol was added to the bath to keep its volume in a constant amount.

For Ni-SiC coating a Ni plate with high purity grade or electrode Ni in Ti basket was applied as anode. There are also different metals which can be used as cathode, such as stainless steel, low-carbon steel, copper, brass, titanium, platinum, and cast-iron.

Before starting plating process, it must be noted that metals with capability of being passive, have to be activated as cathode. Stainless steel, titanium, and aluminum are involved in this. Since graphite can highly reduce the rate of deposition on the surface, cast-iron's surface must be activated, if there is any graphite. Thorough surface activation with 1 molar $PdCl_2$ solution deposited SiC content will increase from 10-20 volumetric percent to 80 volumetric percent.

4.6.2. Other Additives

In order to improve SiC deposition with nickel and to contribute SiC particles to be charged, some other additives, as well as those mentioned, are necessary. Most practical ones for simultaneous deposition of Ni-SiC are some surfactants including: Cetyl Trimethyl Ammonium Bromide (CTAB), Sodium Dodecyl Sulphate (SDS), Azobenzene Trimethyl Ammonium Bromide (AZTAB), saccharin, and sodium hexa-nitro cobalt.

Surfactants are composed of two parts: polar and non-polar part. For this reason they are soluble in both aqueous and organic solvents. One of these two parts is hydrophilic while the other one is hydrophobic. This compatibility of surfactants can change particles hydrophilic behavior to hydrophobic one. Due to their increase in surface charge, cationic surfactants prevent particles to be agglomerated within the electrolyte. This leads to an increase in suspension stability. Besides, increasing surface charge on suspended particles enhance inserted gravitational force from cathode to the particle. So, the content of particles into the coating may have an increase.

These additives are added to electrolyte in very small concentrations. However, their concentration increase up to a definite value generally causes increase of SiC content into the coating. Particles tendency to deposition would be possible through a strong bond between surfactants and particle surface. Nevertheless, as surfactant amount increases, these materials separation from surface will be somehow irregular and the particles prevent SiC deposition through remaining on the surface. This causes SiC content of the coating to decrease.

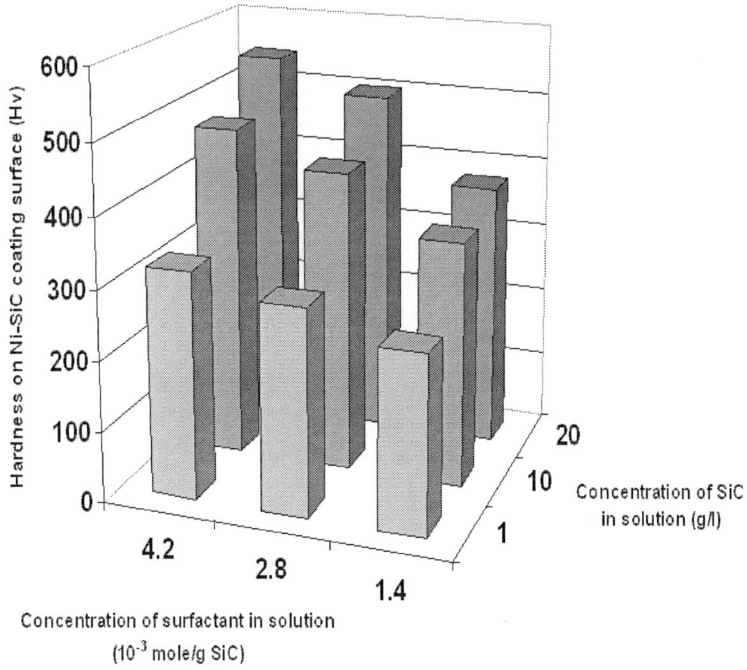

Figure 4.10. Hardness changes with increase of SiC in electrolyte and surfactant concentration

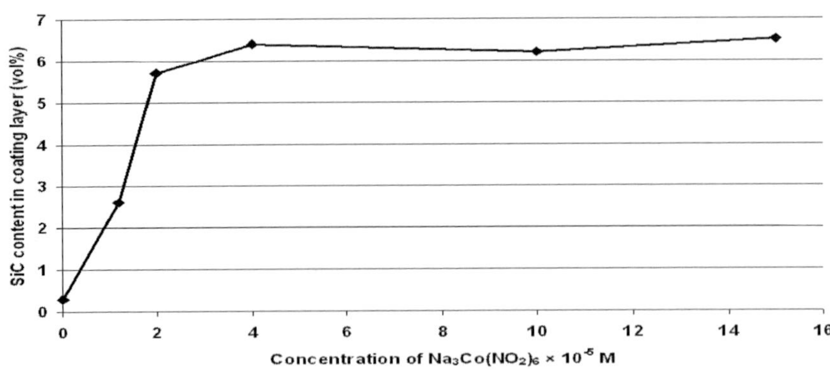

Figure 4.11. Effect of sodium hexa-nitro cobalt increase on coatings SiC content.

Maximum of deposited SiC is varied with every surfactant and is defined through its hydrophobic behavior. For example, increasing alkyl branch length in AZTAB its hydrophobic behavior enhances. Also C4-AZTAB has stronger hydrophobic behavior than that of C2-AZTAB and C0-AZTAB; then segregation rate of these molecules around the cathode during deposition would be as following:

C0-AZTAB > C2-AZTAB > C4-AZTAB

As a result the most volumetric ratio of SiC will be different using each above-mentioned surfactant:

with C0-AZTAB: 71.5 vol%

with C2-AZTAB: 4 vol%

with C4-AZTAB: 50 vol%

As it can be deducted from the curve in figure 4.10, in presence of surfactants active surface molecules entrapment into developing coating, as well as increase of particle ratio within the coating, conducts in enhancement of coating hardness and abrasion resistance. Also, adding a small amount of sodium hexa-nitro cobalt to coating leads to a quick increase of particle ratio into the coating. Here, trivalent cobalt complex leads to increase of SiC ratio and decrease of agglomeration rate within the solution (figure 4.11).

4.6.3. Current

4.6.3.1. Current Density

According previously performed studies increase of current density causes decrease of SiC content of the coating. However, there are some records about increase of SiC particles into the coating with current density increase. Also, there recorded an optimum current density, where values lower than optimum current density causes particles amount while values higher than optimum current density conducts in particles number.

With respect to suggested mechanisms, about simultaneous deposition of particle-metal it can be said that: at high current densities Ni solved ions quickly move from anode toward cathode. However, SiC larger particles which move through convectional mechanism, cannot reach to cathode surface at this speed. As a result the higher amount of Ni will be reduced against particles on the surface and content of particles within the surface will drop. On the other hand, at low current densities, Ni ions are slowly moving into the electrolyte thus there is no enough time for Ni cations to be absorbed. This leads to lack of particles deposition on cathode surface. SiC particles agglomeration at low densities is also due to this low speed of Ni cations movements and lack of full coverage of SiC particles surface with them.

On the other hand, an unusual increase of current density conducts in increase of coatings internal stresses and, as a consequence, SiC particles rejection. With respect to all discussed points it seems that the required current density for obtaining maximum particle volumetric percentage into the coating has an optimum value. It may be suggested this optimum current density is influenced by some other factors such as bath agitation rate, size and amount of particles of the bath. However, there is no clear record for its amount and it just can be told the studies are generally performed using current of 1 to 10 A/dm^2.

4.6.3.2. Current Type

Direct, pulse, and triangular currents are applied for producing composite coating. Changing the current type has some effects on SiC content of the coating and the other features of the coating, which all will be discussed separately. The amount of coating SiC produced under pulse current, particularly at low duty cycle, is higher than that of direct current. Figure 4.12 illustrates SiC ratio changes into the coating with applied current's duty cycle for SiC nano and micro particles. However the ratio of nano-particles into the coating is of less quantity, these particles would be a greater number than micro-particles. Figure 4.13 demonstrates SiC ratio of coating at different concentrations of the particles into the electrolyte.

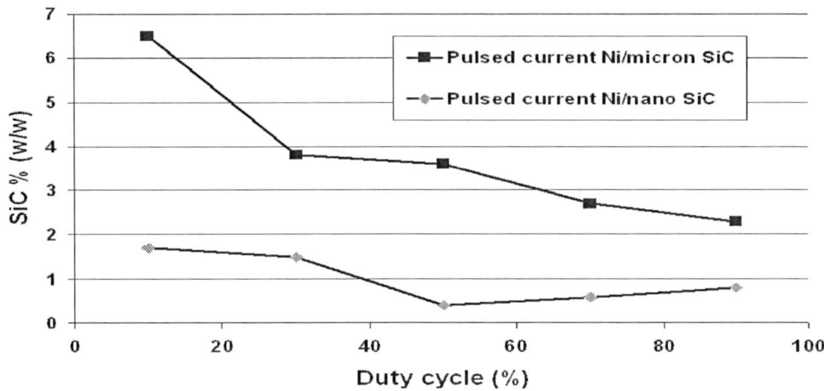

Figure 4.12. Effect of current type on concentration of coatings' SiC nano and micro particles.

Also, composite coating has higher hardness under triangular and pulse current than that of direct current. This increase in hardness with current type changing from direct to pulse is true for both nano and micro particles (figure 4.14). Under pulse current, Ni nucleation rate increases, as well as increase of SiC content of the coating, and Ni matrix particle size of the coating will reduce.

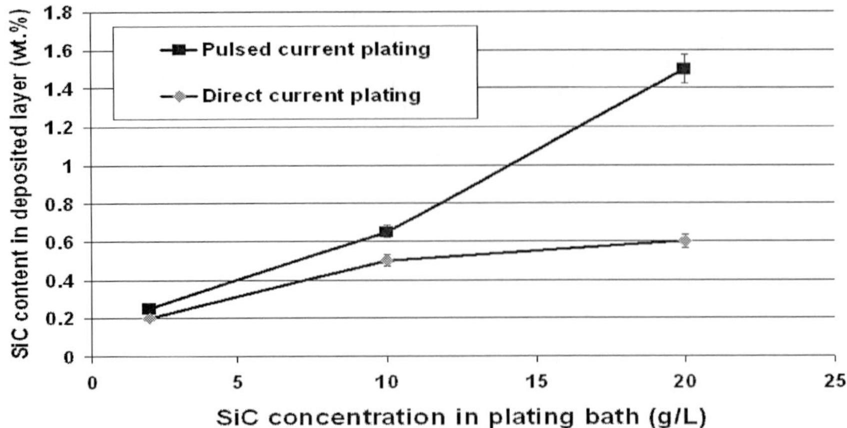

Figure 4.13. Effect of current type on SiC ratio into the coating with SiC concentration changes in the electrolyte.

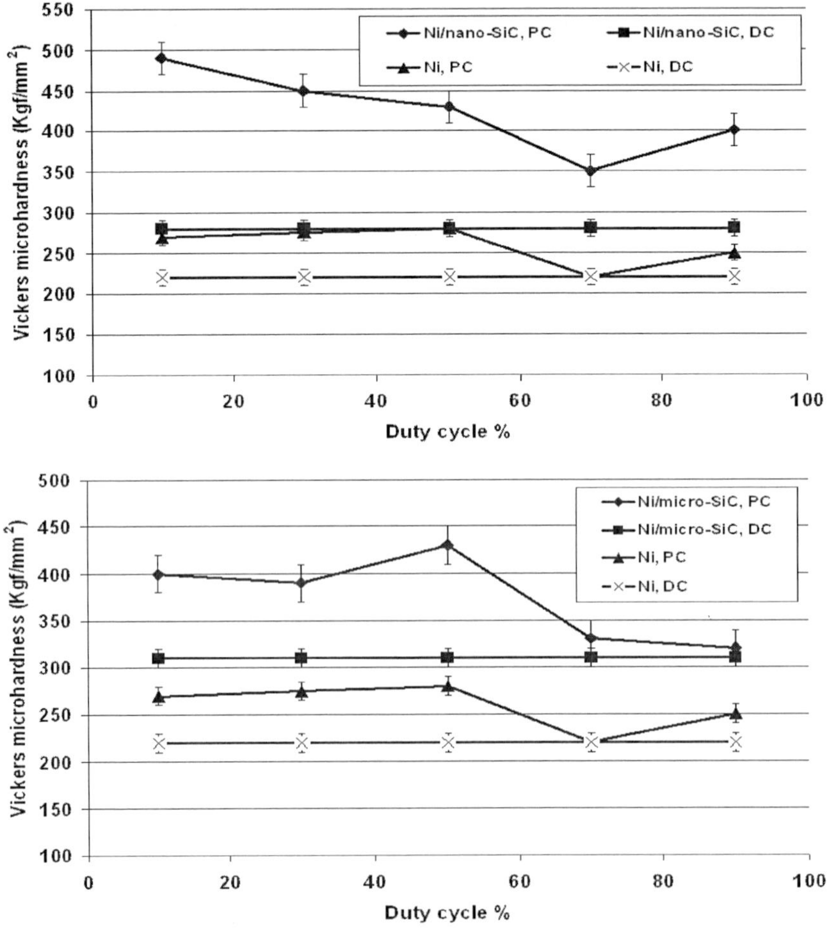

Figure 4.14. Hardness changes with current type and nano and micro particles application.

Whenever the current in pulse state is off and other exterior particles are placed in the coating, they stop its developing when the current is on, again. Hence, at later pulses of the current, Ni will be forced to nucleate. As a result there will be developed a very fine grained coating with high internal stress. Such a fine-grained coating has an extremely high hardness. Once the pause time is longer the coating is more fine-graded. On the other hand application of pulse current encourage Ni coating to change its orientation, [211]. As it previously discussed, such a structural change will noticeably increase Ni coating's hardness and strength. Regarding above-mentioned points it is expected that coating's abrasion resistance under pulse current would have a greater increase, which is turned out to be true in practice. Also, change of current type has a great effect on coating's abrasion resistance.

4.6.4. SiC Particles

The effect of SiC particles into the solution on amount of deposited SiC into the coating can be investigated with particles dimension and SiC content in the solution. Based on performed studies, increase of particle concentration in the electrolyte leads to its increase in number in a coating with a particular particle size. This can be easily justified with Guglielmi's model. Nevertheless, results of more precise studies on effect of particle concentration in electrolyte predict an optimum value. At this optimum concentration the highest number of particles will be placed in the coating (figure 4.15). Increase of coatings particles ratio with electrolyte concentration seems to be somehow inevitable. Once electrolyte concentration passed the optimum value, electrolyte particles will increase to an extent where nickel ions are not able to cover total surfaces of all particles. Then, some particles will have no chance to be deposited. Besides, an extraordinary increase of the concentration brings about particles agglomeration in the electrolyte, due to increase of the chance of particles collision and decrease of their surface charge. Agglomeration sweeps the chance of some particles' deposition and increases agglomerated particles resistance against deposition.

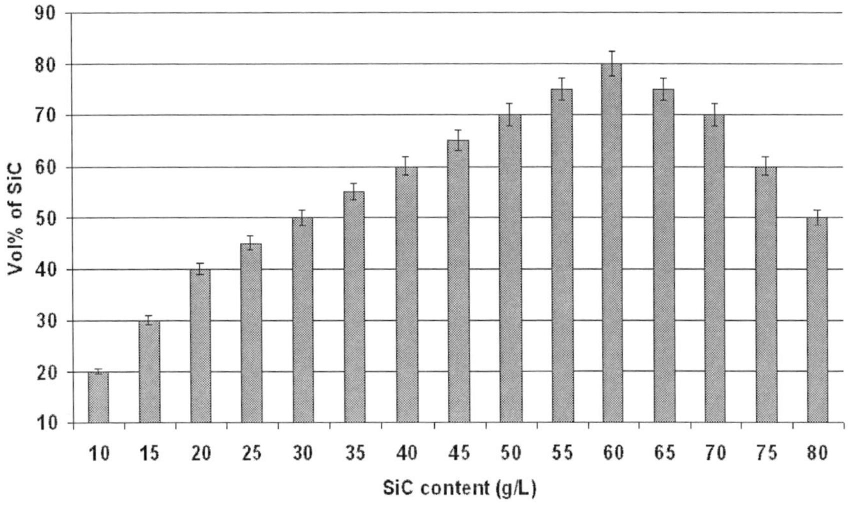

Figure 4.15. SiC ratio changes in coating with its concentration changes in electrolyte.

On the other hand, agglomerated particles, through placing at particles surface proximity, prevent the particles flux toward cathode surface, which in turn decreases SiC ratio of the coating. The optimum concentration of the particles in the electrolyte also depends on particles' size. Then, at this point, it is not possible to declare a particular number.

As previously discussed about particles dimension in the coating, increase of particles size leads to increase of volumetric percentage; but the number of coating particles will decrease. Here, the only noteworthy point is stoppage of coating particles number increase with increase of its dimension behind an optimum value (figure 4.16). As the particles size increases, the amount of absorbed Ni ions on their surface will have an enhancement, and then there is a higher Columbus force applied on the particle. This in turn causes to higher and stronger absorption on cathode surface. Furthermore, through greater increase of particles dimensions weight increase overweighs to Columbus Force increase and particles will settle down after hitting with cathode, which leads to decrease of coating's particles ratio.

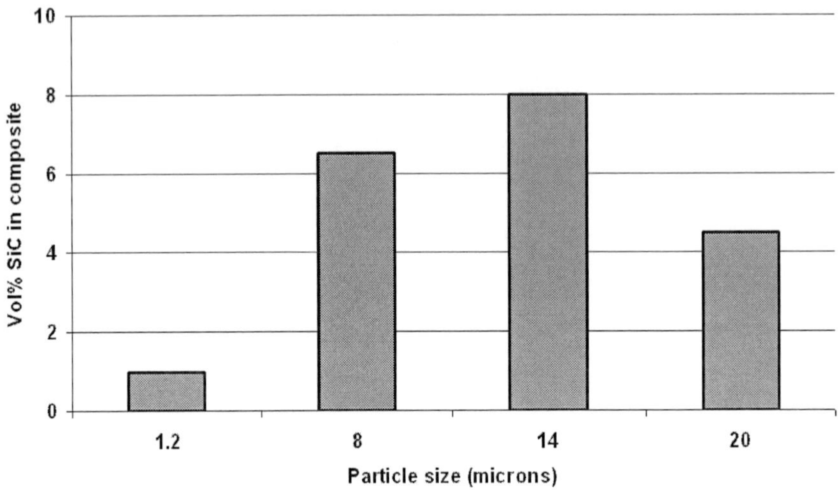

Figure 4.16. Effect of SiC particles size on its simultaneous deposition rate.

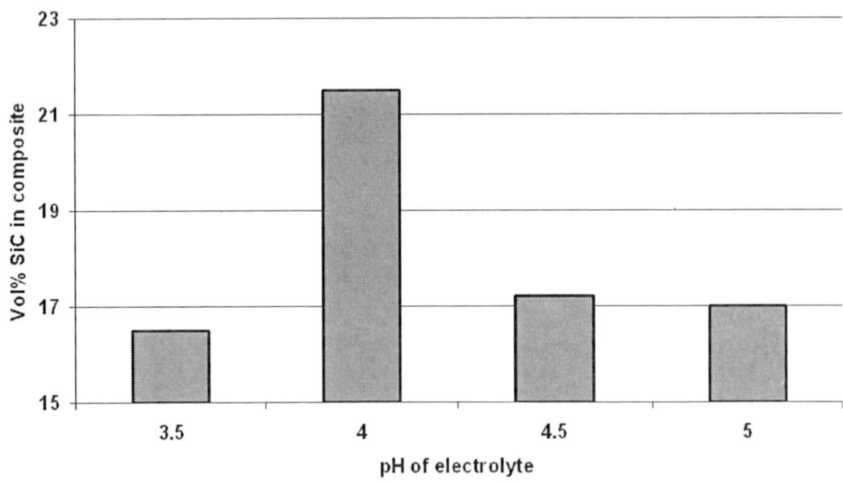

Figure 4.17. pH effects on SiC particles deposition.

4.6.5. pH

pH of electrolyte is another important factor on Ni-SiC's simultaneous deposition. It must be avoided to work at pHs lower than 3 in sulphamate baths since at these pHs composition of nickel sulphamate changes into nickel ammonium sulphate (which is of a low solubility at aqueous environment). In addition, and based on researchers records, working with pHs lower than 3 in Watts baths also dramatically decreases SiC amount of the coating. Produced coatings at high pHs are opaque and crashed with insufficient cohesion. Through different studies, pH of 4 offers the best results. Iso Electric Point (IEP) point for SiC is about pH=3. Then, it is expected to have negative surface potentials at suggested pHs, but contrarily the surface has a positive potential through absorbing Ni^{++} ions. With potential increase, particles electrophoretic mobility and particle density at cathode surface proximity will have an increase. Then, the particles chance of placing in the coating will increase, which consequently enhances the number of SiC in the coating. So, it seems that the used bath composition has a role on optimum pH value. However, regarding applied pHs with different researchers, it seems that pHs of 3.5 to 5 have an efficient yield for Ni-SiC simultaneous deposition.

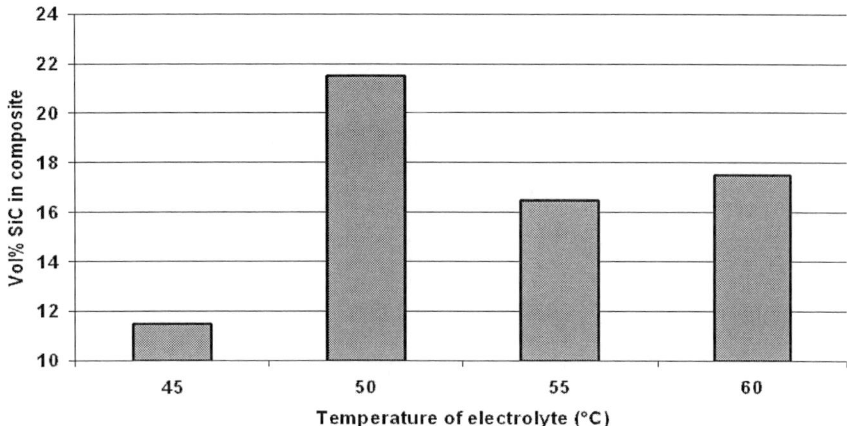

Figure 4.18. Temperature effects on SiC deposition.

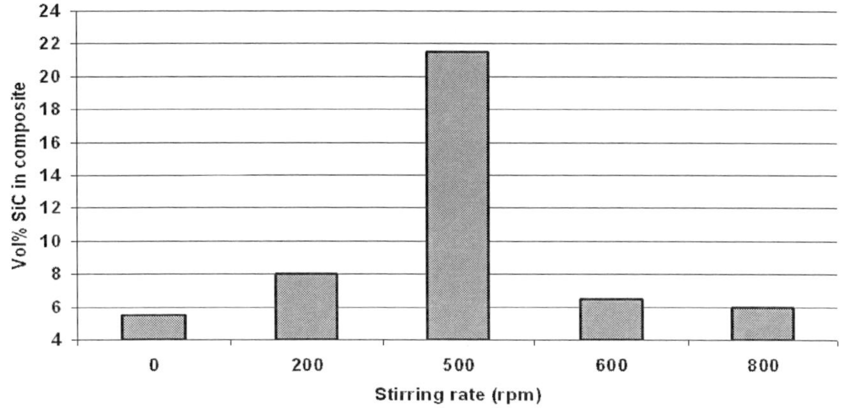

Figure 4.19. Agitation speed effects on SiC deposition.

4.6.6. Temperature

Temperature of 50°C is most common used one among researchers. Also, study results about temperature effect on amount of deposited SiC prove this temperature (figure 4.18). The higher increase of temperature will decrease coatings' SiC content. The reason for drop of coating particles number may be reduce of effective current in electrolyte with temperature increase.

4.6.7. Agitation

Electrolyte agitation during simultaneous deposition process is performed to maintain particles in suspended state in the solution and prevent its precipitation. Besides, particles should reach to cathode surface to deposit. Agitation causes particles to hit with cathode and as a result particle content of the coating will have an increase. On the other hand unusual increase of agitation rate can lead to intense collision of particles and electrolyte with cathode surface, which in turn may conduct in separation of surface absorbed particles. Thus, this causes coating's particle content, as well as particles lodging from developing coating (figure 4.19)

4.7. Simultaneous Deposition Mechanism

Different researchers have offered variant theories for mechanism of insoluble particles' simultaneous deposition in metallic matrix. Some other groups suggested that the reason for simultaneous deposition is particles movement through electrophoretic force toward cathode and their entrapment in developing metallic coating. According this theory hydrogen ion has a key role on simultaneous deposition. On the other hand, some other researchers believe that particles movement toward cathode is done by inserted mechanical force from electrolyte, due to the turbulence created by electrolyte. Results of accurate studies show that neither electrolyte nor mechanical force by itself is satisfying reason for developing of composite coating. Indeed it can be said that both mentioned mechanisms cooperatively provide required conditions for particles movement toward cathode and consequently their entrapment in developing metallic coating.

So many researchers worked on offering a model to predict and explain particles simultaneous deposition in metallic matrix. Theoretical models are focused on deposition mechanisms and prediction of particles deposition amount in composite. Most models do not predict a change in metal reduction rate and method. Guglielmi is among those who tried for first time to offer a mathematical model for particle-metal simultaneous deposition phenomenon. From then on, so many models have been offered. Among them just four models are widely used by researchers. These four models include: Guglielmi model, MTM, Valdes model, and trajectory model. In following section Guglielmi model, as well as a brief explanation of the other three models, will precisely discussed - for its usage to explain and justify Ni-SiC deposition process.

4.7.1. Guglielmi Model

Guglielmi explains a two-staged model which is obtained from simultaneous combining of electrophorus effects and physical absorption. The model is based on two successive absorption steps. At first step, absorption is basically physical which produce a layer of weakly absorbed particles on the surface. At the second step, the absorption is raised from electrical field at proximity of the electrode, and then it has a chemical nature which can absorb the particles on the surface with a great power and intensity. These particles will be gradually buried under metallic matrix and develop a composite.

It worth to tell at first step neutral particles are yet buried with a layer composed of ions and solvent, and this is practically one which is subjected to electrode and particle's interaction. At second step the existed electrical field contributes to surface to be cleaned off from ions and solution.

Through this model weak absorption coating (σ) and strong absorption coating (θ) is defined as following:

$\theta = S_s/S$ and $\sigma = S_1/S$

where, S, S_1, and S_s are electrode surface, surface of weakly absorbed sub-particles, and surface of strongly absorbed sub-particles, respectively. The relationship between σ and C (total concentration of suspended particles in electrolyte) can be explained through Longmir absorption isotherm. Here, the only difference will be a coefficient of (1-θ), for calculating actual available surface for the particles. Thus, we will have:

$$\sigma = 1 + \frac{kC}{1+kC}(1-\theta)$$

where, k is a constant which intensely depend on interaction of electrode and particles. The current model suggested the exponential function of $v0e^{B\eta}$ for this process dependency to electrical field. Where, η is over-potential of metal deposition and v_0 and B are constants.

Hence, strong absorption rate can be defined as follows:

$$dV_p/dt = v_0 e^{\beta\eta}(1-\theta)\frac{kC}{1+kC}$$

At above equation dV_p/dt is the volume of strong absorbed particles at each $1cm^2$ of electrode surface in 1 second.

Volumetric rate of metal precipitation on the surface can be easily obtained by Faraday equation:

$$dV_m/dt = \frac{Wi}{nFd}$$

where, F, W, d, and I are Faraday constant, atomic weight of metal, and current density, respectively.

Once particles concentration is defined as volumetric ratio (α), it can be easily shown that overall deposition (particle + metal) rate will be as following:

$$dV_m/dt = \frac{dV_m}{dt(1-\alpha)} = \frac{Wi}{nFd(1-\alpha)}$$

On the other hand current density has a relationship with η. For approximating great field the equation will change into Tafel equation:

$$dV_m/dt = \frac{dV_m}{dt(1-\alpha)} = \frac{Wi}{nFd(1-\alpha)}$$

where, (1- θ) is used to supply actual available surface for the deposition. Then:

$$dV/dt = \frac{W(1-\theta)i_0 e^{A\eta}}{nFd(1-\alpha)}$$

At stable condition α can be written as ratio of strong absorption rate over total deposition speed:

$$\alpha = \frac{dV_p/dt}{dV/dt}$$

Different arrangement will produce the following general equation, which explains relationship between α, particle concentration in electrolyte, and over-potential.

$$\frac{\alpha}{1-\alpha} = \frac{nFdv_0}{Wi_0} e^{(B-A)\eta} \frac{kC}{1+kC}$$

V_0 and B are particle-dependent parameters, which have same role as i and A in metal deposition.

Usually α is very small value, hence term (1 - α) can be emitted from above equation. Then with reversing the equation and its rearranging we will have:

$$\frac{C}{\alpha} = \frac{Wi_0}{nFdV_0} e^{(A-B)\eta} .(C + \frac{1}{k})$$

If left hand of the above equation is drawn in term of C, it will yield some straight group of line for different values of η, which cut C axis at 1/k. these lines gradient will be same as the equation which is placed out of brackets.

$$tg\phi = \frac{Wi_0}{nFdv_0} e^{(A-B)\eta}$$

Often constant current, for its easy way of control, is used instead of over-potential. Considering α is a very small value, we can use $i/i_0 \approx e^{A\eta}$, then we can have:

$$i_0.e^{(A-B).\eta} = i_0.e^{A\eta(1-B/A)} = i_0^{(B/A)}.i_c^{(1-B/A)}$$

Substituting above equations, we will have:

$$\log tg\phi = \log \frac{Wi_0^{B/A}}{nFdv_0} + (1 - \frac{B}{A})\log i$$

Above equation demonstrates the linear relationship between log tgφ and log i. Drawing tgφ in term of I, in coordination system produces a flat line with gradient of (1 – B/A). Hence, required factors will be obtained.

4.7.2. MTM Model

For some incompatibilities of Guglielmi's model, MTM model was introduced. It can be claimed that this model explains Guglielmi's deposition process in another view. The model is based on two basic principles, including:

1. As soon as particle entering into electrolyte or during its primary processing there will develop an ionic cloud around it.
2. Some parts of this ionic cloud must be reduced for particle's absorption and entrapment.

Corssing its route from electrolyte to cathode surface, the particle will pass the following five steps:

1. Ionic species absorption on the particle surface
2. Particle movement under convectional force up to hydrodynamic layer.
3. Particle penetration from cathode's dual layer.
4. Particle absorption, as well as its ionic cloud, on cathode surface.
5. Reduction of some ions part and final irreversible absorption of the particle in the coating.

Assuming models' electrolyte stability and uniformity, we start to examine infiltrated layers' events. Assuming k numbers of total K ions are absorbed on surface for particle entrapment and using probability rules, the model predicts the number of entrapped particles - indeed particles concentration.

$$W\% = \frac{4\pi r^3 \rho \frac{c_p^*}{c_{ion}^*} (\frac{i_{tr}}{i})^\alpha HP_{(k/K)}}{\frac{3Mi}{nF} + 4\pi r^3 \rho \frac{c_p^*}{c_{ion}^*} (\frac{i_{tr}}{i})^\alpha HP_{(k/K)}} .100$$

In above equation, W/O is the weight ratio of entrapped particles, assuming that the particles are spherical. The other parameters include:

C_p: number of particles, C_{ion}: number of ions in $1m^3$ electrolyte, H: hydrodynamic effects coefficient, i_{tr}: transition current form state of charge transmitting to diffusional control, α: the number of absorbed and free ions interaction during current apply, ρ_p: particles density, and P: the probability of particles entrapment. This model is used to examine Cu-Al_2O_3 simultaneous deposition from acidic-sulphate bath, and Au-Al_2O_3 from KCN, and has potential of being applied in other cases.

4.7.3. Valdes Model

Explaining precipitation mechanism, Valdes used concept of complete precipitation form nano-electrochemical systems, which synthetically suggests very quick deposition. Complete precipitation model assumes that all entered particles will be entrapped within a limited distance from surface in instant irreversible way. In RDE model dispersive colloidal particles movements in the electrolyte is performed with keeping trifle mass balance. This results the following equation for particles numerical concentration:

$$(\partial n/\partial t) + (\partial j/\partial r) = 0$$

Also for particles J flux vector, including diffusion and convection, we have:

$$j = -D_p(\partial n/\partial r) + U.n$$

where, U is the convectional speed, caused by all inserted forces and momentums on the particle.

Valdes improved his model with focusing on particles electrochemical reduction nature. In this model electrochemical reduction of electro-active species on particle provides required cohesion. Then, the most important factor of over-potential particles precipitation will be charge transition. Considering proposed kinetic, Valdes proposed following electrochemical equation for simultaneous precipitation speed:

$$i_p = k^0 C_s [\exp(\frac{\alpha zF}{RT}\eta_\alpha) - \exp(-\frac{(1-\alpha)zF}{RT}\eta_\alpha)]$$

where,

K^0: standard constant of electrochemical reaction speed, Cs: absorbed electro-active species concentration on particle surface, and η_a: over-potential of charge transfer. This model is efficiently corresponded with empirical results. For example it predicts that through increase of density current first the amount of deposited particles increases and then after reaching a maximum amount it will decrease.

4.7.4. Trajectory Model

Inserted forces on a particle, which discussed in Valdes model, are investigated through trajectory model. This model is based on understanding electrical fields around electrode, with respect to effective forces on a particle. Trajectory of one particle is achievable trough equation of motion. Once particles connection criterion is supposed to be its contact with surface, it will be possible to calculate deposition rate. Effective forces on particle include:

1. Applied forces by fluid on particles during its convectional motions and their movement in the fluid.
2. Independent applied forces on particle such as gravitational force, buoyancy, and other volumetric forces.

Volumetric flux of colloidal particles with RDE is measured for particles flux in a finite path, which is divided into input particles and rejected particles. Finite trajectory of particles can be calculated in simplest way through Integral of path equations of particles which only hit with electrode.

References

[1] Abdel Aal, A. and Hassan, H.B. Electrodeposited nanocomposite coatings for fuel cell application. *Journal of Alloys and Compounds*, 2009, 477(1-2), 652-656.

[2] Aliofkhazraei, M., Ahangarani, S. and Sabour Rouhaghdam, A. Effect of the duty cycle of pulsed current on nanocomposite layers formed by pulsed electrodeposition. *Rare Metals*, 2010, 29(2), 209-213.

[3] Aliofkhazraei, M. and Sabour Rouhaghdam, A. Fabrication of TiC/WC ultra hard nanocomposite layers by plasma electrolysis and study of its characteristics. *Surface and Coatings Technology*, 2010.

[4] Allain, E., Besson, S., Durand, C., Moreau, M., Gacoin, T. and Boilot, J.P. Transparent mesoporous nanocomposite films for self-cleaning applications. *Advanced Functional Materials*, 2007, 17(4), 549-554.

[5] Aouadi, S.M., Debessai, M. and Filip, P. Zirconium nitride/silver nanocomposite structures for biomedical applications. *Journal of Vacuum Science and Technology B: Microelectronics and Nanometer Structures,* 2004, 22(3), 1134-1140.

[6] Aouadi, S.M., Luster, B., Kohli, P., Muratore, C. and Voevodin, A.A. Progress in the development of adaptive nitride-based coatings for high temperature tribological applications. *Surface and Coatings Technology*, 2009, 204(6-7), 962-968.

[7] Aouadi, S.M., Paudel, Y., Luster, B., Stadler, S., Kohli, P., Muratore, C., Hager, C. and Voevodin, A.A. Adaptive Mo2N/MoS2/Ag tribological nanocomposite coatings for aerospace applications. *Tribology Letters*, 2008, 29(2), 95-103.

[8] Bakhshi, R., Edirisinghe, M.J., Darbyshire, A., Ahmad, Z. and Seifalian, A.M. Electrohydrodynamic jetting behaviour of polyhedral oligomeric silsesquioxane nanocomposite. *Journal of Biomaterials Applications*, 2009, 23(4), 293-309.

[9] Balazs, D.J., Hossain, M.M., Brombacher, E., Fortunato, G., K¶rner, E. and Hegemann, D. Multi-functional nanocomposite plasma coatings - Enabling new applications in biomaterials. *Plasma Processes and Polymers,* 2007, 4(SUPPL.1), S380-S385.

[10] Bauer, F., Ernst, H., Hirsch, D., Naumov, S., Pelzing, M., Sauerland, V. and Mehnert, R. Preparation of scratch and abrasion resistant polymeric nanocomposites by monomer grafting onto nanoparticles, 5a: Application of mass spectroscopy and atomic force microscopy to the characterization of silane-modified silica surface. *Macromolecular Chemistry and Physics*, 2004, 205(12), 1587-1593.

[11] Bauer, F. and Mehnert, R. UV curable acrylate nanocomposites: Properties and applications. *Journal of Polymer Research*, 2005, 12(6), 483-491.

[12] Branagan, D.J., Breitsameter, M., Meacham, B.E. and Belashchenko, V. High-performance nanoscale composite coatings for boiler applications. *Journal of Thermal Spray Technology*, 2005, 14(2), 196-204.

[13] Chandra, A., Turng, L.S., Gopalan, P., Rowell, R.M. and Gong, S. Semitransparent poly(styrene-r-maleic anhydride)/ alumina nanocomposites for optical applications. *Journal of Applied Polymer Science,* 2007, 105(5), 2728-2736.

[14] Chang, M.H., Hsieh, T.E., Huang, B.R., Hsieh, H.E., Juang, F.S., Tsai, Y.S., Liu, M.O. and Lin, J.L. A study of ultraviolet-curable organic/inorganic hybrid nanocomposites and their encapsulating applications for organic light-emitting diodes. *Materials Chemistry and Physics,* 2009, 115(2-3), 541-546.

[15] Chattopadhyay, D.K. and Raju, K.V.S.N. Structural engineering of polyurethane coatings for high performance applications. *Progress in Polymer Science* (Oxford), 2007, 32(3), 352-418.

[16] Chen, W.C. and Lin, Y.Y. Glucose biosensor based on dextran-Fe_3O_4 nanocomposite modified SPCEs. *Biomedical Engineering - Applications, Basis and Communications*, 2009, 21(6), 437-440.

[17] Czyzniewski, A. and Precht, W. Deposition and some properties of nanocrystalline, nanocomposite and amorphous carbon-based coatings for tribological applications. *Journal of Materials Processing Technology,* 2004, 157-158(SPEC. ISS.), 274-283.

[18] De Hosson, J.T.M., Pei, Y. and Cheu, C. Jerky-type phenomena at nanocomposite surfaces: The breakdown of the Coulomb friction law. JOM, 2007, 59(7), 45-49.

[19] Dearnley, P.A., Kern, E. and Dahm, K.L. Wear response of crystalline nanocomposite and glassy Al_2O_3-SiC coatings subjected to simulated piston ring/cylinder wall tests. Proceedings of the Institution of Mechanical Engineers, Part L: *Journal of Materials: Design and Applications,* 2005, 219(2), 121-137.

[20] Devaux, E., Koncar, V., Kim, B., Campagne, C., Roux, C., Rochery, M. and Saihi, D. Processing and characterization of conductive yarns by coating or bulk treatment for smart textile applications. *Transactions of the Institute of Measurement and Control,* 2007, 29(3-4), 355-376.

[21] Duan, L., Xie, J., Zhang, D., Wang, L., Dong, G., Qiao, J. and Qiu, Y. Nanocomposite thin film based on ytterbium fluoride and N/N²-Bisi(1-naphthyl)-N,N-diphenyl-1,1-Biphenyl-4,4- diamine and its application in organic light emitting diodes as hole transport layer. *Journal of Physical Chemistry* C, 2008, 112(31), 11985-11990.

[22] Eklund, P. Novel ceramic Ti-Si-C nanocomposite coatings for electrical contact applications. *Surface Engineering*, 2007, 23(6), 406-411.

[23] Es-Souni, M., Kartopu, G., Habouti, S., Piorra, A., Es-Souni, M., Solterbeck, C.H. and Brandies, F.H. Processing and thin film formation of TiO_2-Pt nanocomposites. *Physica Status Solidi (A) Applications and Materials,* 2008, 205(2), 305-310.

[24] Fang, F.F., Choi, H.J. and Joo, J. Conducting polymer/clay nanocomposites and their applications. *Journal of Nanoscience and Nanotechnology*, 2008, 8(4), 1559-1581.

[25] Fang, W.C., Chyan, O., Sun, C.L., Wu, C.T., Chen, C.P., Chen, K.H., Chen, L.C. and Huang, J.H. Arrayed CNx NT-RuO_2 nanocomposites directly grown on Ti-buffered Si substrate for supercapacitor applications. *Electrochemistry Communications*, 2007, 9(2), 239-244.

[26] Fernando, R. Nanomaterial technology applications in coatings. *JCT CoatingsTech,* 2004, 1(5), 32-38.

[27] Franchi, D. and Rostagno, M. Nanocomposite thin coatings for wear control: Applications and limits. Rivestimenti sottili nanostrutturati per il controllo dell'usura: *Applicazioni e limiti*, 2005, 97(11-12), 21-26.

[28] Fujii, S., Aichi, A., Akamatsu, K., Nawafune, H. and Nakamura, Y. One-step synthesis of polypyrrole-coated silver nanocomposite particles and their application as a coloured particulate emulsifier. *Journal of Materials Chemistry,* 2007, 17(36), 3777-3779.

[29] Ganjaee, M., Mohseni, M., Mohajerani, E., Aghili, Y. and Moradian, S. Preparation of organically modified hybrid nanocomposites for optical applications. *Journal of Optoelectronics and Advanced Materials*, 2008, 10(3), 578-588.

[30] Ginder, J.M., DePinto, J.T., Cooper, S., Gotcher, A., Doll, G.L., Rardon, D. and Verbrugge, M.W. Nanotechnology in automotive applications. *Advanced Materials and Processes*, 2005, 163(12), 25-28.

[31] Gorokovsky, V.I., Bowman, C., Gannon, P.E., VanVorous, D., Voevodin, A.A., Muratore, C., Kang, Y.S. and Hu, J.J. Deposition and characterization of hybrid filtered arc/magnetron multilayer nanocomposite cermet coatings for advanced tribological applications. *Wear*, 2008, 265(5-6), 741-755.

[32] Hamdy, A.S. Novel approaches in designing high performance nano and nano-composite coatings for industrial applications. *International Journal of Nanomanufacturing*, 2009, 4(1-4), 235-241.

[33] Hanisch, C., Kulkarni, A., Zaporojtchenko, V. and Faupel, F. Polymer-metal nanocomposites with 2-dimensional au nanoparticle arrays for sensoric applications. *Journal of Physics: Conference Series*, 2008, 100(PART 5).

[34] Hou, X. and Choy, K.L. Processing and applications of aerosol-assisted chemical vapor deposition. *Chemical Vapor Deposition*, 2006, 12(10), 583-596.

[35] Iyer, G., Pinaud, F., Tsay, J., Li, J.J., Bentolila, L.A., Michalet, X. and Weiss, S. Peptide coated quantum dots for biological applications. *IEEE Transactions on Nanobioscience,* 2006, 5(4), 231-238.

[36] Japelj, B., Vuk, A.S., Orel, B., Per[illegible]e, L.S., Jerman, I. and Kova[illegible], J. Preparation of a TiMEMO nanocomposite by the sol-gel method and its application in coloured

thickness insensitive spectrally selective (TISS) coatings. *Solar Energy Materials and Solar Cells*, 2008, 92(9), 1149-1161.

[37] Jiang, L.Y., Wu, X.L., Guo, Y.G. and Wan, L.J. SnO_2-based hierarchical nanomicrostructures: Facile synthesis and their applications in gas sensors and lithium-ion batteries. *Journal of Physical Chemistry C,* 2009, 113(32), 14213-14219.

[38] Jiang, W., Malshe, A.P. and Goforth, R.C. Cubic Boron Nitride (cBN) based nanocomposite coatings on cutting inserts with chip breakers for hard turning applications. *Surface and Coatings Technology*, 2005, 200(5-6), 1849-1854.

[39] Joshi, M., Thakare, V. and Pal, S.K. Polymer clay nanocomposites for coated textile applications: A perspective. Man-Made Textiles in India, 2005, 48(12), 448-455.

[40] Kamalan Kirubaharan, A.M., Selvaraj, M., Maruthan, K. and Jeyakumar, D. Synthesis and characterization of nanosized titanium dioxide and silicon dioxide for corrosion resistance applications. *Journal of Coatings Technology Research,* 2009, 1-8.

[41] Kang, T.J., Yoon, J.W., Kim, D.I., Kum, S.S., Huh, Y.H., Hahn, J.H., Moon, S.H., Lee, H.Y. and Kim, Y.H. Sandwich-type laminated nanocomposites developed by selective dip-coating of carbon nanotubes. *Advanced Materials*, 2007, 19(3), 427-432.

[42] Khazrayie, M.A. and Aghdam, A.R.S. Si_3N_4/Ni nanocomposite formed by electroplating: Effect of average size of nanoparticulates. *Transactions of Nonferrous Metals Society of China*, 2010, 20(6), 1017-1023.

[43] Koo, J.H. and Pilato, L.A. Polymer nanostructured materials for high temperature applications. *SAMPE Journal*, 2005, 41(2), 7-19.

[44] Kulisch, W., Colpo, P., Gibson, P.N., Ceccone, G., Shtansky, D.V., Levashov, E.A., Jelinek, M., Philip, P.J.M. and Rossi, F. Hybrid ICP/sputter deposition of TiC/CaO nanocomposite films for biomedical application. *Applied Physics A: Materials Science and Processing*, 2006, 82(3 SPEC. ISS.), 503-507.

[45] Kwong, C.Y., Choy, W.C.H., Djuri, A.B., Chui, P.C., Cheng, K.W. and Chan, W.K. Poly(3-hexylthiophene): TiO_2 nanocomposites for solar cell applications. *Nanotechnology*, 2004, 15(9), 1156-1161.

[46] Lang, S., Beck, T., Schattke, A., Uhlaq, C. and Dinia, A. Characterization of nanostructured coatings based on oxides for tribological applications. *Surface and Coatings Technology*, 2004, 180-181, 85-89.

[47] Lewin, E., Olsson, E., Andr, B., Joelsson, T., Berg, Wiklund, U., Ljungcrantz, H. and Jansson, U. Industrialisation study of nanocomposite ncTiC/a-C coatings for electrical contact applications. *Plasma Processes and Polymers*, 2009, 6(SUPPL. 1), S928-S934.

[48] Lewin, E., Olsson, E., Andr, B., Joelsson, T., Berg, Wiklund, U., Ljungcrantz, H. and Jansson, U. Erratum: Industrialisation study of nanocomposite nc-TiC/a-C coatings for electrical contact applications (Plasma Processes and Polymers (2009) 6 (S928-S934) DOI: 10.1002/ppap.200932303). *Plasma Processes and Polymers*, 2009, 6(SUPPL. 1).

[49] Lewin, E., Wilhelmsson, O. and Jansson, U. Nanocomposite nc-TiC/a-C thin films for electrical contact applications. *Journal of Applied Physics*, 2006, 100(5).

[50] Li, T.C., G³ˡes, M.S., Fabregat-Santiago, F., Bisquert, J., Bueno, P.R., Praslttichal, C., Hupp, J.T. and Marks, T.J. Surface passivation of nanoporous TiO2 via atomic layer deposition of ZrO_2 for solid-state dye-sensitized solar cell applications. *Journal of Physical Chemistry* C, 2009, 113(42), 18385-18390.

[51] Louro, C., Lamni, R. and Levy, F. W-B-N sputter-deposited thin films for mechanical application. *Surface and Coatings Technology*, 2005, 200(1-4 SPEC. ISS.), 753-759.

[52] Lu, S.W. and Schmidt, H.K. Nanostructures, optical properties, and imaging application of lead-sulfide nanocomposite coatings. *International Journal of Applied Ceramic Technology*, 2004, 1(3), 119-128.

[53] Luo, X.Y., Zuo, D.W., Wang, M., Li, S.L. and Chang, H. Preparation and hot-dip galvanizing application of CeO_2/Zn nanocomposite. *Transactions of Nonferrous Metals Society of China* (English Edition), 2005, 15(SPEC. ISS. 3), 203-207.

[54] Ma, S.B., Nam, K.W., Yoon, W.S., Yang, X.Q., Ahn, K.Y., Oh, K.H. and Kim, K.B. Electrochemical properties of manganese oxide coated onto carbon nanotubes for energy-storage applications. *Journal of Power Sources,* 2008, 178(1), 483-489.

[55] Martucci, A., Buso, D., De Monte, M., Guglielmi, M., Cantalini, C. and Sada, C. Nanostructured sol-gel silica thin films doped with NiO and SnO2 for gas sensing applications. *Journal of Materials Chemistry*, 2004, 14(19), 2889-2895.

[56] Mauermann, M., Eschenhagen, U., Bley, T. and Majschak, J.P. Surface modifications - Application potential for the reduction of cleaning costs in the food processing industry. *Trends in Food Science and Technology*, 2009, 20(SUPPL. 1), S9-S15.

[57] Mosher, B.P., Wu, C., Sun, T. and Zeng, T. Particle-reinforced water-based organic-inorganic nanocomposite coatings for tailored applications. *Journal of Non-Crystalline Solids*, 2006, 352(30-31), 3295-3301.

[58] Muratore, C., Hu, J.J. and Voevodin, A.A. Adaptive nanocomposite coatings with a titanium nitride diffusion barrier mask for high-temperature tribological applications. *Thin Solid Films*, 2007, 515(7-8), 3638-3643.

[59] Murday, J.S., Cotter, F., Belk, J.H., Rawal, S. and Dastoor, M.N. Nanotechnology in aerospace applications. *Advanced Materials and Processes*, 2005, 163(12), 21-23.

[60] Musat, V., Fortunato, E., Botelho do Rego, A.M. and Monteiro, R. Sol-gel cobalt oxide-silica nanocomposite thin films for gas sensing applications. *Thin Solid Films*, 2008, 516(7), 1499-1502.

[61] Necula, B.S., Fratila-Apachitei, L.E., Berkani, A., Apachitei, I. and Duszczyk, J. Enrichment of anodic MgO layers with Ag nanoparticles for biomedical applications. *Journal of Materials Science: Materials in Medicine,* 2009, 20(1), 339-345.

[62] Nobel, M.L., Mendes, E. and Picken, S.J. Acrylic-based nanocomposite resins for coating applications. *Journal of Applied Polymer Science*, 2007, 104(4), 2146-2156.

[63] Nobel, M.L., Picken, S.J. and Mendes, E. Waterborne nanocomposite resins for automotive coating applications. *Progress in Organic Coatings*, 2007, 58(2-3), 96-104.

[64] Nose, M., Kawabata, T., Ohi, M., Nagae, T., Masa, S., Hatano, Y., Ikeno, S. and Nogi, K. Development of PVD - TiAlN/a-C nanocomposite coating for pressure die casting applications. *Plasma Processes and Polymers*, 2007, 4(SUPPL.1), S681-S686.

[65] Nol, S., Alamarguy, D., Houz, F., Benedetto, A., Viel, P., Palacin, S., Izard, N. and Chenevier, P. Nanocomposite thin films for surface protection in electrical contact applications. *IEEE Transactions on Components and Packaging Technologies,* 2009, 32(2), 358-364.

[66] Oey, C.C., Djurii, A.B., Kwong, C.Y., Cheung, C.H., Chan, W.K., Nunzi, J.M. and Chui, P.C. Nanocomposite hole injection layer for organic device applications. *Thin Solid Films,* 2005, 492(1-2), 253-258.

[67] Pang, X. and Zhitomirsky, I. Electrodeposition of nanocomposite organic-inorganic coatings for biomedical applications. *International Journal of Nanoscience,* 2005, 4(3), 409-418.

[68] Papaefthymiou, G.C., Rabias, I., Fardis, M., Devlin, E., Boukos, N., Tsitrouli, D. and Papavassiliou, G. Gummic acid stabilized Fe_2O_3 aqueous suspensions for biomedical applications. *Hyperfine Interactions*, 2009, 190(1-3), 59-66.
[69] Park, B.J., Sung, J.H., Park, J.H., Choi, J.S. and Choi, H.J. Poly(vinyl acetate)/clay nanocomposite materials for organic thin film transistor application. *Journal of Nanoscience and Nanotechnology*, 2008, 8(5), 2676-2679.
[70] Pei, Y.T., Galvan, D., De Hosson, J.T.M. and Cavaleiro, A. Nanostructured TiC/a-C coatings for low friction and wear resistant applications. *Surface and Coatings Technology*, 2005, 198(1-3 SPEC. ISS.), 44-50.
[71] Polcar, T., Vitu, T., Cvrcek, L., Novak, R., Vyskocil, J. and Cavaleiro, A. Tribological behaviour of nanostructured Ti-C:H coatings for biomedical applications. *Solid State Sciences*, 2009, 11(10), 1757-1761.
[72] Prasad, S.V., Scharf, T.W., Kotula, P.G., Michael, J.R. and Christenson, T.R. Application of diamond-like nanocomposite tribological coatings on LIGA microsystem parts. *Journal of Microelectromechanical Systems,* 2009, 18(3), 695-704.
[73] Rao, Y. and Wong, C.P. Material Characterization of a High-Dielectric-Constant Polymer-Ceramic Composite for Embedded Capacitor for RF Applications. *Journal of Applied Polymer Science*, 2004, 92(4), 2228-2231.
[74] Ruland, W. and Smarsly, B.M. Two-dimensional small-angle X-ray scattering of self-assembled nanocomposite films with oriented arrays of spheres: Determination of lattice type, preferred orientation, deformation and imperfection. *Journal of Applied Crystallography*, 2007, 40(3), 409-417.
[75] Samouhos, S. and McKinley, G. Carbon nanotube-magnetite composites, with applications to developing unique magnetorheological fluids. Journal of Fluids Engineering, *Transactions of the ASME*, 2007, 129(4), 429-437.
[76] Scharf, T.W., Prasad, S.V., Dugger, M.T., Kotula, P.G., Goeke, R.S. and Grubbs, R.K. Growth, structure, and tribological behavior of atomic layer-deposited tungsten disulphide solid lubricant coatings with applications to MEMS. *Acta Materialia*, 2006, 54(18), 4731-4743.
[77] See, S.C., Zhang, Z.Y. and Richardson, M.O.W. A study of water absorption characteristics of a novel nano-gelcoat for marine application. *Progress in Organic Coatings*, 2009, 65(2), 169-174.
[78] Sharma, S., Soni, V.P. and Bellare, J.R. Electrophoretic deposition of nanobiocomposites for orthopedic applications: Influence of current density and coating duration. *Journal of Materials Science: Materials in Medicine,* 2009, 20(SUPPL. 1), S93-S100.
[79] Shu, C.H., Chiang, H.C., Tsiang, R.C.C., Liu, T.J. and Wu, J.J. Synthesis of organic-inorganic hybrid polymeric nanocomposites for the hard coat application. *Journal of Applied Polymer Science*, 2007, 103(6), 3985-3993.
[80] Silvain, J.F., Vincent, C., Heintz, J.M. and Chandra, N. Novel processing and characterization of Cu/CNF nanocomposite for high thermal conductivity applications. *Composites Science and Technology*, 2009, 69(14), 2474-2484.
[81] Singh, M.K., Shokuhfar, T., De Almeida Gracio, J.J., De Sousa, A.C.M., Da Fonte Fereira, J.M., Garmestani, H. and Ahzi, S. Hydroxyapatite modified with carbon-nanotube-reinforced poly(methyl methacrylate): A nanocomposite material for biomedical applications. *Advanced Functional Materials,* 2008, 18(5), 694-700.

[82] Sudo, A., Uenishi, K. and Endo, T. Imidazole-promoted copolymerization of epoxide and 3,4-dihydrocoumarin and its application to a high-performance curing system. *Journal of Polymer Science, Part A: Polymer Chemistry*, 2007, 45(16), 3798-3802.

[83] Takafuji, M., Ide, S., Ihara, H. and Xu, Z. Preparation of poly(1-vinylimidazole)-grafted magnetic nanoparticles and their application for removal of metal ions. *Chemistry of Materials*, 2004, 16(10), 1977-1983.

[84] Tanaka, T., Montanari, G.C. and Mulhaupt, R. Polymer nanocomposites as dielectrics and electrical insulation- perspectives for processing technologies, material characterization and future applications. *IEEE Transactions on Dielectrics and Electrical Insulation*, 2004, 11(5), 763-784.

[85] Teyssier, J., Le Dantec, R., Galez, C., Mugnier, Y., Bouillot, J. and Plenet, J.C. $LiIO_3/SiO_2$ nanocomposite for quadratic non-linear optical applications. *Journal of Non-Crystalline Solids*, 2004, 341(1-3), 152-156.

[86] Vandevyver, E. and Eichholz, E. Latest advancements in PVC/clay nanocomposites: Potential applications in plastisols. *Plastics, Rubber and Composites,* 2008, 37(9-10), 417-420.

[87] Veprek, S. Reproducibility of deposition and industrial applications of stable superhard nanocomposites. *Transactions of Nonferrous Metals Society of China* (English Edition), 2004, 14(SUPPL. 2), 19-24.

[88] Veprek, S. and Veprek-Heijman, M.J.G. Industrial applications of superhard nanocomposite coatings. *Surface and Coatings Technology*, 2008, 202(21), 5063-5073.

[89] Vida-Simiti, I., Jumate, N., Negrea, G., Sechel, N. and Coman, C. Structure of some composite materials for excessive wear applications. *Metalurgia International*, 2009, 14(SPEC. ISS. 3), 137-140.

[90] Voevodin, A.A. and Zabinski, J.S. Nanocomposite and nanostructured tribological materials for space applications. *Composites Science and Technology,* 2005, 65(5 SPEC. ISS.), 741-748.

[91] Wang, T.L., Hwang, W.S. and Yeh, M.H. Preparation, properties, and anticorrosion application of poly(methyl methacrylate)/montmorillonite nanocomposites coating on brass via solution polymerization. *Journal of Applied Polymer Science,* 2007, 104(6), 4135-4143.

[92] Wei, H., Liu, J., Zhou, L., Li, J., Jiang, X., Kang, J., Yang, X., Dong, S. and Wang, E. $[Ru(bpy)3]^{2+}$-doped silica nanoparticles within layer-by-layer biomolecular coatings and their application as a biocompatible electrochemiluminescent tag material. *Chemistry* (Weinheim an der Bergstrasse, Germany), 2008, 14(12), 3687-3693.

[93] Yao, S.H., Su, Y.L., Kao, W.H. and Cheng, K.W. Evaluation on wear behavior of Cr-Ag-N and Cr-W-N PVD nanocomposite coatings using two different types of tribometer. *Surface and Coatings Technology*, 2006, 201(6), 2520-2526.

[94] Yi, D.K. Synthesis and applications of crack-free SiO_2 monolith containing CdSe/ZnS quantum dots as passive lighting sources. *Journal of Nanoscience and Nanotechnology,* 2008, 8(9), 4538-4542.

[95] Yu, Y.H., Yeh, J.M., Liou, S.J., Chen, C.L., Liaw, D.J. and Lu, H.Y. Preparation and properties of polyimide-clay nanocomposite materials for anticorrosion application. *Journal of Applied Polymer Science*, 2004, 92(6), 3573-3582.

[96] Yudin, V.E., Otaigbe, J.U., Gladchenko, S., Olson, B.G., Nazarenko, S., Korytkova, E.N. and Gusarov, V.V. New polyimide nanocomposites based on silicate type nanotubes: Dispersion, processing and properties. *Polymer,* 2007, 48(5), 1306-1315.

[97] Zeng, Z., Zhou, Y., Zhang, B., Sun, Y. and Zhang, J. Designed fabrication of hard Cr{single bond}Cr_2O_3{single bond}Cr_7C_3 nanocomposite coatings for anti-wear application. *Acta Materialia*, 2009, 57(18), 5342-5347.

[98] Zhang, F., Yan, P.X., Chen, J.T., Miao, B.B., Li, G.B. and Wang, J. Recent research and applications of $(Ti_{1-x}Al_x)N$ thin hard coating. Rengong Jingti Xuebao/*Journal of Synthetic Crystals*, 2007, 36(3), 699-704.

[99] Zhang, L., Wang, W., Shang, M., Sun, S. and Xu, J. Bi_2WO_6@carbon/Fe_3O_4 microspheres: Preparation, growth mechanism and application in water treatment. *Journal of Hazardous Materials,* 2009, 172(2-3), 1193-1197.

[100] Zhang, X., Yang, J., Zeng, Z., Huang, L., Chen, Y. and Wang, H. Stabilized dispersions of titania nanoparticles via a sol-gel process and applications in UV-curable hybrid systems. *Polymer International*, 2006, 55(4), 466-472.

[101] Zhou, Q., Cha, J.H., Huang, Y., Zhang, R., Cao, W. and Shung, K.K. Alumina/epoxy nanocomposite matching layers for high-frequency ultrasound transducer application. *IEEE Transactions on Ultrasonics, Ferroelectrics, and Frequency Control*, 2009, 56(1), 213-219.

Chapter 5

Corrosion Properties of Nanocomposite Coatings

Abstract

Electrochemical properties of nanocomposite coatings and some examples were discussed in this chapter. It was focused on metallic matrix composites in the examples. In nanocomposite coatings, the size of reinforced particles has reduced from several micrometers to under 100 nm, so this causes a highly increase in hardness and strength, coating erosive and corrosion resistance to the great extent. Achieving to such favorable properties have led to widely use of nanocomposite coatings in different industries such as manufacture of cutting and abrasive tools, military industries etc. The amount of improvement in properties of this kind of composites depends on reinforced particles kind, their size, percent and way of propagation in metallic matrix.

5.1. Introduction

Electrodeposited metallic matrix composite coatings, because of having favorable properties like erosive and corrosive resistance, magnetic properties for application in chemical, automotive, military industries and manufacture of parts of microsystems etc. have been widely regarded. In 1928, the initial idea for application of composite coatings was raised in order to create self-lubricating coating in car engines. In 1966, coatings with nickel matrix comprising of silicon carbide were used for industrial application in manufacture of rotary engines which was the first industrial application for composite coatings. During recent years, global research view has returned into using these coatings with nanostructure and or presence of nanoparticles. This causes hardness, strength, and erosive and corrosive resistance are increased remarkably in coating. Making alloyed of nickel coatings with alloyed elements like cobalt, tungsten may noticeably improve its hardness, strength, ductility and thermal stability, and this has caused taking an approach toward using composite coatings with nickel alloyed matrix. On the other hand, reduced grain size in nanometric quantities may highly increase whether in coating hardness and strength, but it reduces coating ductility and adhesion. Thus, many of world researches during recent years about production of nanostructured coatings have been focused in improvement of the given properties. The results have shown that application of alloyed elements and adjustment of matrix microstructure and favorable propagation of ceramic particles through basic methods are solutions for the aforesaid problems.

5.2. Introducing of Composite Coatings

By entering particles like TiO_2, Al_2O_3, SiO_2 and carbides like SiC, WC, and nitrides such as Si_3N_4 and diamond or even polymeric particles into metallic matrix or alloy, metallic matrix composite coatings are produced. During recent years, such coatings have been widely used within different industries in terms of appropriate properties like resistance against erosion, corrosion, oxidation, and wear etc.

It is possible to create composite coatings through different methods like electrodeposition, electrolysis, CVD, PVD, PACVD etc. Because of its ease of application and cheapness, low temperature process, ease of access to nanostructure as well as production of coating with high density and pore- free, electrodeposition is one of the appropriate techniques for application of such coatings and it has been highly appreciated by the researchers.

Nickel based coatings are some of coatings which are highly regarded due to having favorable properties like resistance against erosion and corrosion, magnetic properties etc. Along industry current advancement, pure nickel coatings could not meet the existing requirements. For this reason, several efforts were made in order to adjust nickel coating properties by different techniques. Of these methods, one could refer to creation of a secondary phase in the form of diffused non- metallic particles in the matrix which are called nickel matrix composite coatings. These coatings are often created by using ceramic micrometric particles. In recent years, global researches approach has been oriented toward application of such coatings by using nanometric particles.

In nanocomposite coatings, the size of reinforced particles has reduced from several micrometers to under 100 nm, so this causes a highly increase in hardness and strength, coating erosive and corrosive resistance to the great extent. Achieving to such favorable properties have led to widely use of nanocomposite coatings in different industries such as manufacture of cutting and abrasive tools, military industries etc. The amount of improvement in properties of this kind of composites depends on reinforced particles kind, their size, percent and way of propagation in metallic matrix.

5.3. Application of Composite Coatings

During recent years, nanocomposite materials have been widely used in chemical, nuclear, automotive industries etc, due to hardness and highly strength, and resistance against erosion and corrosion, and excellent oxidation and thermal high stability. Now, this category of materials has been used in manufacture of hydrophobic anodes, photo catalyst and photo electro catalyst materials. Application of such materials is also addressed in manufacture of *Micro electro Mechanical Systems* (MEMS) and *High Aspect Ratio Micromachining* (HARM). HARM microsystems have about several micrometers height with about a few micrometers to tens micrometer width and they are made from materials like metal, ceramic and polymer. Due to their wide area (ranged from tens centimeter to several square meters), these materials are used in different fields like heat transfer, fluid mechanics, composite materials, bearings and catalyst systems.

Because of erosive resistance, hardness and high corrosive resistance, nanocomposite materials show that there are highly susceptible for manufacture of microsystems next generations. To manufacture microsystem devices, the used composite material should have particles in nanometric dimensions. Similarly, nanocomposite coatings have been applied to increase lifetime in cylinder and piston components as well as car exhaust.

5.4. Electrodeposition Process in Production of Nanocomposite Coatings

Electrodeposition of composite coatings includes some type of electroplating where some particles in micron or less than micron dimensions are suspended in electrolyte and they enter into coating during the coating process. Electrodeposition in manufacture of composite coatings includes electroplating and electrolysis techniques. In electrolysis method, metallic coating deposit are generated on the surface of piecework by chemical reactions, and the dominant mechanism is to particles strong attraction which covered by reduction of appropriate quantity of metallic complex ions at the level of existing particles over cathode surface. However in electroplating, metallic coating is deposited by using external current and reduction of metallic ions appropriate quantity over cathode surface and the main mechanism is strong absorption of reduced particles which are absorbed weakly by cathode. Of electrolysis method advantages, one could imply further hardness and uniformity, and of electrodeposition technique advantages to further ductility and electrolyte lower cost and more speed in coating process etc.

One could use direct and pulsed current in production of nanocomposite coatings by electroplating technique. In pulsed electroplating, one may use different forms like triangular, saw tooth, and squared pulses etc. In pulsed electroplating technique, greater amount of particles with higher uniformity enter into the matrix than in direct electroplating, and on the other hand, due to presence of more parameters in production process, it is possible to control coating properties in this technique. Production parameters like current density and operational cycle, frequency, solution revolving velocity and additive have highly affect on the quantity of strengthening particles as well as their propagation in the matrix.

5.5. Effective Factors on Coating Microstructures Produced by Electroplating

Coating microstructures produced by electroplating play important role in coating mechanical properties and corrosion. Therefore, it is very important to control microstructure in production of such materials. In electroplating, several parameters are effective in microstructure like current type and its specifications, electrolyte temperature and pH, additive materials to electrolyte etc. Following of this subject, we refer to some of effective parameters on coating microstructures which are produced by electroplating technique.

Current density: Studies have indicated that over- potential changes may create some changes in morphology, texture and size of grain in coatings which produced by electroplating. In low current density where over- potential rate is low, coating surface is coarser and structure and grain are larger, and by increase in current density, coating surface becomes smoother, roughness is lesser and coating grains are finer. Increasing current density may prevent from preferred growth and reduced unevenness over coating surface.

Similarly, current density may affect on element density of alloyed coatings as well. During investigation into the effect of pulsed cathodic current density on deposited cobalt quantity for Ni-Co it was seen that by increase in current density deposited cobalt quantity is reduced on coating. At this study, the given electrolyte included 3gr/lit saccharin and 0.05gr/lit Sodium lauryl sulfate. By change of current density from 20 to 160A/dm^2, cobalt quantity was changed from 12 to 6% weight percentage on coating. This result is in compliance with the resulted which obtained by other researchers in production of Ni-Co coating through High-speed jet electrodeposition. This is because of the fact that cobalt deposit and nickel deposit are controlled by penetration and activation respectively; as a result, increase in current density may increase electrode activity and reaction and consequently increase in deposited nickel percentage on coating.

In figure 5.1, grain size change is seen in Co-Ni coating by increase in current density. In current density with rate of 20, 100, and 120 A/dm^2, grain size reaches to 160, 20, and 10 nanometers. This is due to increase in nucleation rate by increasing current density. Density of lesser currents leads to colony structures with more clear grain boundary, coarser surfaces and with less compression. Also it is observed that extinction longer times in electrodeposition with pulsed current may lead to bigger grains. Up to 120 A/dm^2 and with reduction of grain size, hardness grows more. By more increase in current density coating hardness lowers; it is because of increase in hydrogen bubbles and or bigger grains and or less efficiency of nickel deposit and all of which lead to increase in tension and porosity in coating thus decrease in hardness.

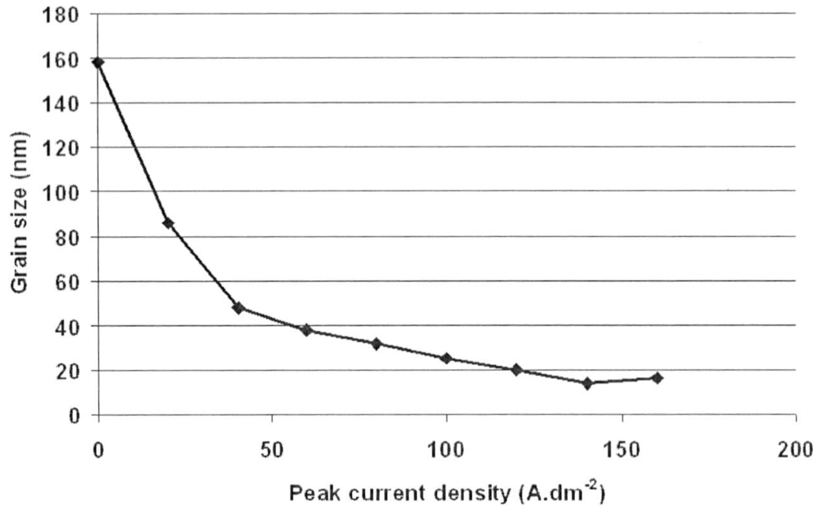

Figure 5.1. Changes in Ni-Co grain size by increase in current density.

Also, current density may affect highly on rate of particles presence in coating. In chloride electrolyte the higher amount of nanoparticles were observed in lower current densities, while in citrate electrolyte in higher current densities, higher density of particles in the coating was obtained. This shows that current density effect varies by change in electrolyte composition. Increase in current density may decrease the presence of diamond particles in coating and increase in coating surface coarseness were observed in electrodeposition of Ni-Diamond and Ni-Co-Diamond composite coatings. Similar results were seen in production of micrometric and nanometric Ni-alumina composite coatings as well as in bronze-graphite nanocomposite coatings. In electrodeposition of Ni-SiC nanocomposite coatings it was seen that initially weight percentage of the given particles increases in coating by reducing current density, and by reaching to a maximum rate and through further reduction of density, weight percentage of particles decreases in the coating. Shrestha observed similar results in production of Ni- Diamond coatings. It was seen Ni-Co-SiC coating electrodeposition that by increase in current density up to 20 A/dm^2, the particles percentage increases because of their growing tendency to enter into this coating. But in the rates greater than 20 A/dm^2 because of faster reduction of metallic ions and increase in their deposits on cathode surface, particles quantity decreases in the coating.

The results obtained from recent studies on the produced coatings by electrodeposition indicate that extinction and on and off time periods as well as their ratio may affect highly on crystallographic texture and matrix morphology and the quantity and propagation of ceramic particles in matrix. Increase in on time within 1-5 milliseconds may lead to decrease in grain size, since it is due to increase in over-potential and thus in nucleation speed. Similarly, increase in on time will lead to change in matrix texture from random distribution to strong fiber texture (200). On the other hand, the longer extinction times may cause coarsening of grain size and changing of crystallographic orientation. Also it was seen that by increase in off time, more particles enter into the coating. Over-increase in on time, of course, may increase the possibility of removal of particles from cathode surface, so this leads to decrease in particle presence in the coating.

Type of current: In order to create a coating by electrodeposition, one may use different currents like square pulsed current (PC), triangular, and pulsed reverse current (PRC). Many studies have shown that application of pulsed current may lead to further presence of nanometric particles in the coating; additionally, it provides possibility of a wider range for composite compounds and their related properties. Electrodeposition by pulsed current may improve hardness and morphology, ductility, surface coarseness as well as lesser porosity and greater resistance against erosion. In electrodeposition with pulsed current, because of this fact that during off periods it prevents from growing of crystalline sheets, on the other hand at the beginning of on time, nucleation process is urged a finer and more uniform structure may be obtained than by direct current. Of course, it was seen in some explorations that particles deposit rate is greater in direct electrodeposition than in pulsed current, and it is likely due to faster burying the given great agglomerates that cause non-uniformity in the structure. Although ceramic particles percentage was greater when using direct current, of course, but due to non- homogeneous propagation of particles in coating, erosive resistance and hardness have been lesser in coatings which produced by pulsed current. This indicates that in addition to ceramic particles weight percentage, homogeneity and fineness of particles propagation also affect highly on coating mechanical properties. Using triangular pulse may produce finer structure due to nucleation greater rate than square pulse in Ni-SiC coatings which created by pulsed current. This shows that pulse shape may affect on particles co-deposition. $Ni-ZrO_2$ nanocomposite coating which produced by PRC current had ceramic particles with greater hardness and erosion lesser rate than coating produced by PC current. Grain size in coating which produced by PRC current was greater than coatings produced by PC current, so this is because of solving small grains during current anodic periods.

Type of anions: Type of anions may also affect highly on coating structures. By adding different salts to the electrolyte of electrodeposition, the given salts are transformed into cations and anions, and the effect of such salts depends on the produced anions and cations. By mixing these salts, a wide range of different morphologies is created. Different anions have various sizes and masses, for instance, chloride anions are light-weight and small and have penetration greater speed toward anode and create a coarse surface, while sulfate anions are greater with more viscosity and create wave form surface. Great anions cause interference in cations movement toward cathode and anions toward anode. Such interference may lead to distributed charge reduction incidence on cathodic surface and creation of smooth surface. In smaller anions, there is no such interference and reduction of metallic ions occurs preferentially on surface beads and creating rough surface. Sulfate baths cause creation of smooth surface but full of pores, while chloride baths may lead creating coarse surface with lesser porosities and with more efficient current. By mixture of chloride and sulfate baths Watts bath is created which has high efficiency for high current, and leads to creating smooth surface without porosities.

Temperature: Increase in temperature may increase grain size and roughness on coating surface. Temperature rise may also increase ions penetration and their mobility as well as average increase in atoms surface permeability which are placed on cathode surface by adsorption, so all of these increase grain growth velocity thus creation of bigger grains with rougher surface. One of the effective parameters on particles quantity and propagation in coating is electrolyte temperature during coating process. For example, in electroplating of Ni-Co-SiC composite coating, it is seen that by temperature rise up to 40°C and due to increase in particles activity, the percentage of particles becomes lesser in coating. For $Ni-Al_2O_3$ nanostructure coating, this applies inversely.

Additive materials: Additives for the bath may also highly affect on microstructure. The common additive which used in fabrication of nanocomposite coatings that produced by electrodeposition usually include grain refiners, surfactants, and salts of electrodeposition alloyed elements like Co, W etc. Grain refiner materials are like saccharin and minerals with heavy molecules which act as growth inhibitors on coating surface and lead to creation of fine microstructures. Surfactants are also the materials which are usually added to coating in electrolyte for adaptation of particles surface charge and improvement in their propagation and quantity. By generation of electrostatic repulsive forces, mineral additives prevent from adhesion of particle with homonymous charges to each other. If using cation surfactants, through creation of surface positive charges on particles, it not only prevents from particles sintering but also will improve electrostatic adsorption of particles on cathode surface (particles weak absorption). Also the next reduction of such materials on cathode surface may help to strong absorption of particle in coating. Application of SDS surfactant (sodium 2-decyl sulfate with $CH_3(CH_2)_{11}OSO_3Na$ chemical formula) in production of Ni-SiC nanocomposite coating extremely decreased agglomerates size and homogeneous propagation and fine particles in the coating. Ger showed that application of Cetyl Trimethyl Ammonium Bromide (CTAB) surfactant in electrodeposition bath for Ni-SiC nanocomposite coating not only decreases agglomerates diameter with finer and more homogeneous propagation of particles in the coating but also particles volumetric percentage becomes further in coating and by which erosive resistance is improved. Sheretha obtained the similar results from the studies on Azobenzene 4-oxyethyl Trimethyl Ammonium Bromide (AZTAB) surfactant on Ni-Diamond coating. In a study, in order to increase alumina particles volume in nickel coating, Hexadecyle Pyridinium Bromide (HPB) surfactant has been used.

Adding ions of alloyed elements to electrolyte may highly affect on morphology and microstructure. Increase of cobalt ion in electrolyte may increase cobalt percent in coating. By implementation of Ni-Co coating by high-speed jet electrodeposition method, it was observed that increase in cobalt ion density may lead dislocation of peaks of X-ray diffraction due to constant increase in crystallographic grid. Similarly, it was seen in this technique that increase in electrolyte agitation and density of cobalt ion and current density may decrease grain size, and increase in density of nickel ion and decrease in electrolyte temperature and current density, and increase in electrolyte agitation may increase cobalt deposit in the coating.

Application of ultrasonic waves: Results of these studies have shown that potential of ultrasonic waves during electrodeposition may lead to increase grain size and reduced presence of particles in the coating and increase in ions permeability and consequently increased speed of deposition. Similarly, this causes removal of bubbles from particles and cathode surface and increasing particles adhesion to cathode surface.

Ultrasonic waves cause reduced velocity of grain growth and grain size by making interruption in grains growth; and increase in ultrasonic waves potential may decrease preferred textures and widening of peaks in X-ray diffraction within Ni-Co-Al_2O_3 coating. Similarly, increase in ultrasonic wave potential may reduce cobalt amount in Ni-Co-Al_2O_3 coating. Propagation of ultrasonic waves in the electrolyte creates very large sectional stresses which are thousands times greater than atmospheric pressure, so this leads to break down the bound among particles in agglomerate. Then some bubbles are created and enter into new levels among particles while preventing particles from sintering.

pH change: pH change in electrodeposition bath causes change in zeta potential of particles surface, so it is effective in particles propagation. On the other hand, studies have shown that bath pH is effective in grain size and thus on coating hardness and strength.

5.6. Exploration of Erosive and Corrosive Properties of Composite Coatings

The main objective in fabrication of nanocomposite coatings is to increase in erosive resistance, hardness and resistance against corrosion. Many results in studies indicate that the presence of ceramic particles in coating may highly increase the resistance against erosion and corrosion. The reason for increasing resistance against erosion is attributed to factors like increase in hardness, and decrease in adhesive erosion between overlapping surfaces and adjustment of microstructure. The important point which should be considered in study of erosive properties of this category of materials is in addition to quantity and propagation of particles, they also play a significant role in resistance against erosion. Fine and homogenous propagation of particles causes improvement in erosive properties.

Some of researchers argue that silicon carbide particles act as nucleation centers themselves and through which they prevent grains from growth during coating growth process and consequently they are converted into fine grain structure with high hardness and resistant against erosion, some others believe in that SiC particles through absorption of hydrogen ions create sectional basic conditions adjacent to coating-electrolyte interface which eventually lead to presence of texture (210) that stronger than more ductile texture (100); in fact, in the absence of silicon carbide in electrolyte, ductile texture (100) is dominant texture in coating which is a texture with high ductility, low hardness and high internal tension. Moreover, this group of researchers believes in that this sectional change in electrolyte composition prevents grain from growth while urging nucleation; so accordingly, by creation of finer grain structure, it contributes to hardness increase trend and coating erosive resistance.

The other reason for increase in coating hardness may be due to the effect of distribution-hardness mechanism so that placed particles on coating act as barriers against dislocations. It should be noticed, of course, that one could benefit from this effect well if non-agglomerate particles are propagated homogeneously and with less than one micron thickness placed on the coating. It was reported that decrease in silicon carbide particles size may affect positively on resistance against erosion and corrosion in Ni-SiC composite coatings. Despite of this fact, a decrease is possible in grains size and may lead to lesser presence of particles in coating.

In study on Ni-SiC composite coatings with micro- and nano-particles, it was seen that although volumetric percent of micrometric particles was greater than coating with silicon carbide nanometric particles, but coating with nanometric particles may indicate greater hardness than its rival coating. This shows that not only volumetric percentage of the placed particles on coating is important, but density of the particles which placed in the coating should also be considered. In some studies, it is implied that erosive resistance of coating with nano-particles is lesser than coatings with micro-particles. This is because of splitting of nano-particles from coating more easily and creation of the related tri-body scratching erosion.

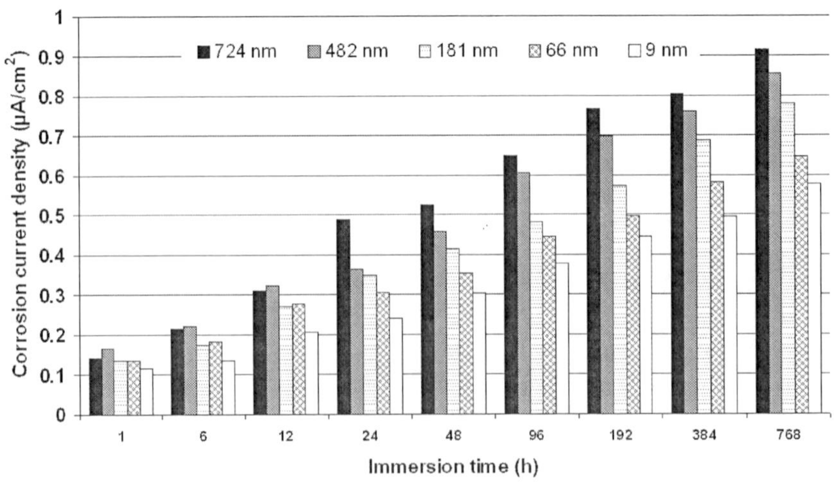

Figure 5.2. Changing trend of corrosion current densities for different ASNPs after immersion in 3.5 wt.% NaCl solution in room temperature.

The exploration results on Ni-SiC nanocomposite coating indicated that the presence of particles on coating surface may noticeably decrease corrosion rate. In an oxygen-enriched atmosphere, Ni-Al_2O_3 and Ni-WC nanocomposite coatings showed better corrosive and oxidation resistance than pure nickel. In the study on corrosive properties of Ni-Al_2O_3 in NaCl solution, it was seen that presence of alumina particles to 6 percent weight in nanocomposite coating may decrease 6 times in corrosion rate than in pure nickel coating. Similarly, the further presence of particles in the coating may increase the resistance against corrosion. During study on corrosive properties on Ni-CeO_2 composite coating, it was characterized that it is more resistant to corrosion than pure nickel coating. This is because of presence of fewer regions which are susceptible to corrosion in composite coating, since a part of the surface has been covered by ceramic particles with higher corrosive resistance.

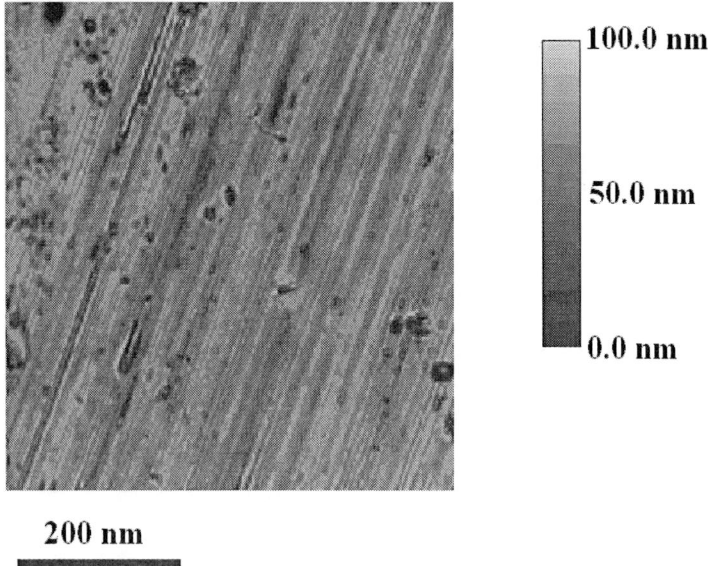

Figure 5.3. AFM nanostructure of nanometric pits of the layer with ASNP equal to 9 nm.

5.7. Examples of Corrosion Improvement by Nanocomposite Coatings

5.7.1. Ni-W/Al_2O_3/CNT Nanocomposite Coating

Figure 5.2 illustrates changing trend of corrosion current densities (CCD) after immersion in 3.5 wt.% NaCl solution in room temperature for Ni-W/Al_2O_3/CNT electrodeposited nanocomposite coatings. As it can be seen from this figure, however CCD will increase a little after long time immersion in corrosive solution but lowering the average size of nanoparticles (ASNP) will lead to lower CCDs which at first indicate that decreasing ASNP is useful for decreasing CCD. However observations with unequipped eye did not indicate presence of pitting phenomena but AFM studies show that nanocomposites with very low ASNP will show some nanometric pits. Figure 5.3 shows the observed pits for the nanocomposite layer with ASNP equal to 9 nm.

5.7.2. Ni-W/CNT Nanocomposite Coating

Plots of tafel polarization tests for Ni-W and nanocomposite layers formed by different duty cycles (or different concentrations of carbon nanotubes in the metallic matrix) are shown in figure 5.4. Changing trend of corrosion potentials (E_{corr}) and corrosion current densities (i_{corr}), which were analyzed through tafel plots, are illustrated in figure 5.5. Addition of nanoparticulates will shift corrosion potentials of electroplated layers towards noble direction (positive values). Increasing the concentration of carbon nanotubes will have an effect in an opposite manner, but the changing amount is ignorable. Changing trend of corrosion current densities shows that increasing the amount of carbon nanotubes has an optimum level for decreasing them. It was found that 9.1 wt.% of carbon nanotubes in the metallic matrix will show minimum corrosion current density of nanocomposite layer. Surfaces have approximately no porosity, so it can be concluded that differences in electrochemical properties are just related to the content of carbon nanotubes and their distribution and amount of agglomeration in the metallic matrix. When carbon nanotubes distribute uniformly, they will protect nanocomposites and substrates from corrosive agents and decrease the corrosion current densities. In an opposite manner, the agglomerated nanoparticulates will decrease the electrochemical properties of the obtained layer.

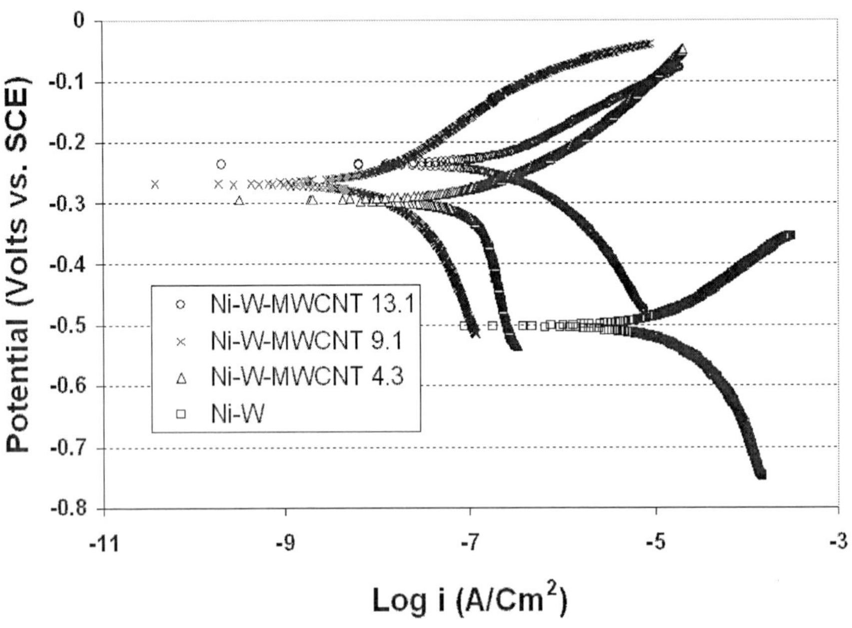

Figure 5.4. Tafel plots of (□) Ni-W and Ni-W/CNT nanocomposite layers formed by different duty cycles of pulsed current ((Δ) 20% of duty cycle (or 4.3 wt.% of carbon nanotubes content), (×) 50% of duty cycle (or 9.1 wt.% of carbon nanotubes content), and (○) 80% of duty cycle (or 13.1 wt.% of carbon nanotubes content)).

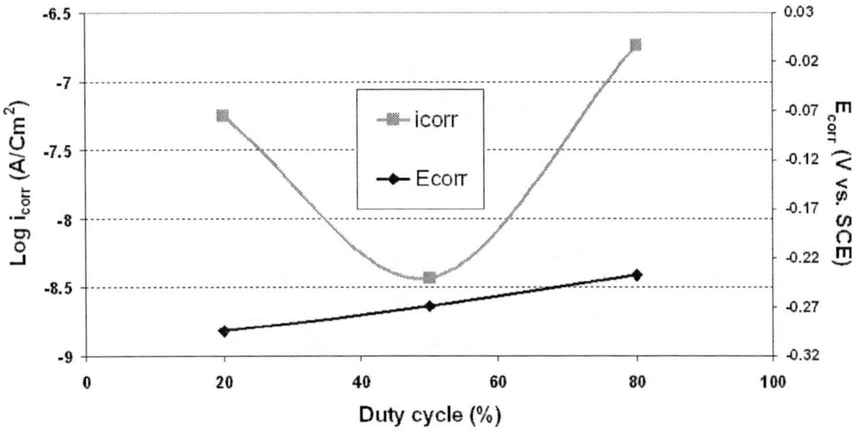

Figure 5.5. Changing trend of E_{corr} and i_{corr} with respect to applied duty cycle (or concentration of CNT nanoparticulates in the metallic matrix).

5.7.3. Ni/Al_2O_3/Y_2O_3/CNT Nanocomposite Coating

The Taguchi method for the design of experiment has been used for optimizing tertiary nanocomposite electrodeposited coating process parameters for the corrosion protection of treated samples. The contribution of Y_2O_3 concentration is more than the sum of the contributions of all the other three factors. It is evident that, among the selected factors, Y_2O_3 concentration has the major influence on the corrosion rate of performed coatings. It can be seen that the current density is second important factor that affects on corrosion rate of the treated substrates. Furthermore, it can be assumed that treatment time and temperature of electrolyte have almost the same effect on corrosion rates of coatings because of the minor difference in the contribution percentages among these two factors. By ranking their relative contributions, the sequence of the four factors affecting the corrosion rate is Y_2O_3 concentration, current density, treatment time and temperature of electrolyte. In the case of average size of nanoparticles ranking of effective factors by their relative contributions is as same as for corrosion rate which show strong relation among these two measured properties of coatings. AFM and TEM analysis have confirmed smooth surface and average size of nanoparticles in the optimal coating.

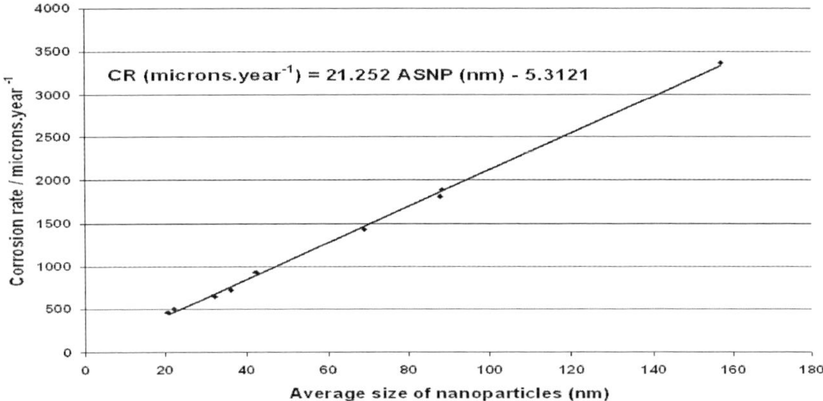

Figure 5.6. Relation among average size of nanoparticles and corrosion rate.

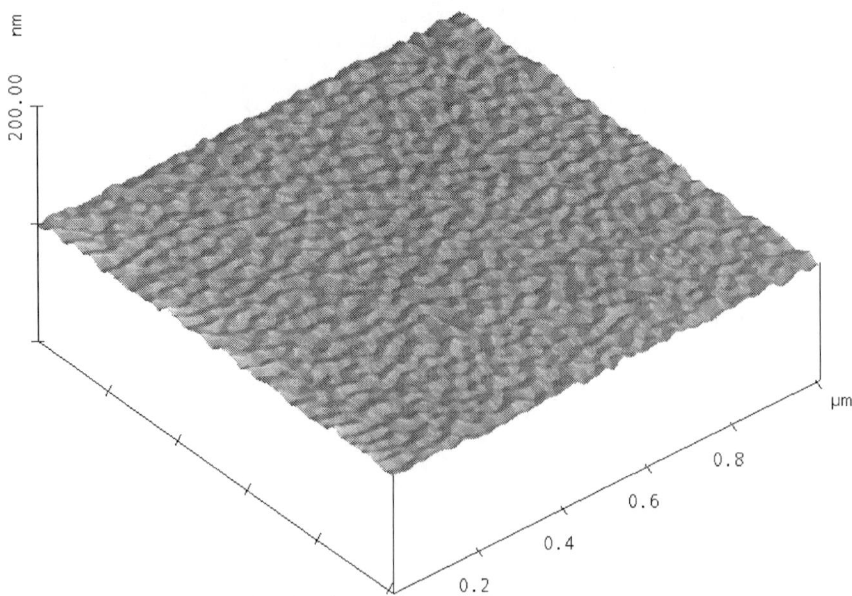

Figure 5.7. AFM surface profile of optimal coating.

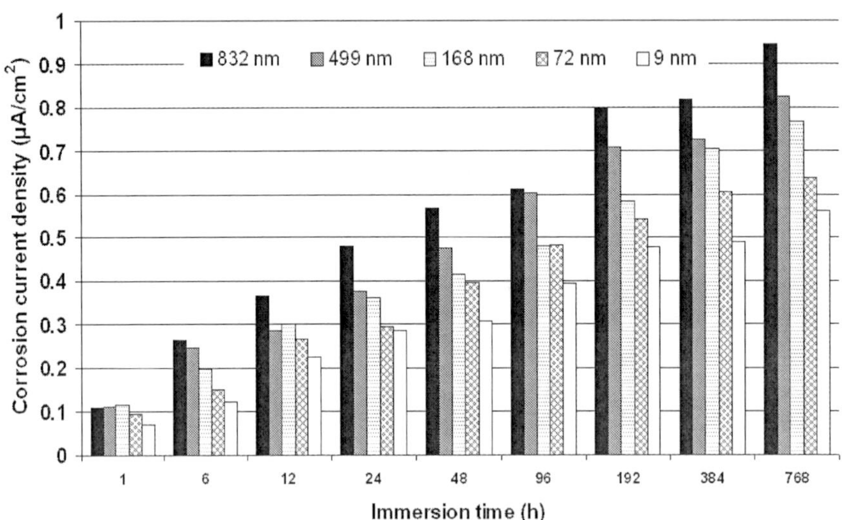

Figure 5.8. Changing trend of corrosion current densities for different ASNPs after immersion in 3.5 wt.% NaCl solution in room temperature.

5.7.4. Ni/Si_3N_4 Nanocomposite Coating

Figure 5.8 illustrates changing trend of corrosion current densities (CCD) after immersion in 3.5 wt.% NaCl solution in room temperature. As it can be seen from this figure, however CCD will increase a little after long time immersion in corrosive solution but lowering the ASNP will lead to lower CCDs which at first indicate that decreasing ASNP is useful for decreasing CCD. However observations with unequipped eye did not indicate presence of pitting phenomena but AFM studies show that nanocomposites with very low ASNP will show some nanometric pits.

REFERENCES

[1] Abdel Aal, A. Hard and corrosion resistant nanocomposite coating for Al alloy. *Materials Science and Engineering* A, 2008, 474(1-2), 181-187.

[2] Alam, J., Riaz, U., Ashraf, S.M. and Ahmad, S. Corrosion-protective performance of nano polyaniline/ferrite dispersed alkyd coatings. *Journal of Coatings Technology Research*, 2008, 5(1), 123-128.

[3] Aliofkhazraei, M., Ahangarani, S. and Sabour Rouhaghdam, A. Effect of the duty cycle of pulsed current on nanocomposite layers formed by pulsed electrodeposition. *Rare Metals*, 2010, 29(2), 209-213.

[4] Aliofkhazraei, M., Ahangarani, S.H. and Rouhaghdam, A.S. Effect of surface nanocrystallization and PPEC time on complex nanocrystalline hard layer fabricated by plasma electrolysis. *Transactions of Nonferrous Metals Society of China*, 2010, 20(3), 425-431.

[5] Aliofkhazraei, M., Hassanzadeh-Tabrizi, S.A., Sabour Rouhaghdam, A. and Heydarzadeh, A. Nanocrystalline ceramic coating on [gamma]-TiAl by bipolar plasma electrolysis (effect of frequency, time and cathodic/anodic duty cycle). *Ceramics International*, 2009, 35(5), 2053-2059.

[6] Aliofkhazraei, M., Morillo, C., Miresmaeili, R. and Sabour Rouhaghdam, A. Carburizing of low-melting-point metals by pulsed nanocrystalline plasma electrolytic carburizing. *Surface and Coatings Technology,* 2008, 202(22-23), 5493-5496.

[7] Aliofkhazraei, M., Rouhaghdam, A.S., Ghobadi, E. and Mohsenian, E. Electrodeposition and mechanical and corrosion resistance properties of tertiary Ni-W/Al_2O_3/CNT nanocomposite coatings. *Advanced Materials Research*, 2010 (89-91), 12-162010).

[8] Aliofkhazraei, M. and Sabour Roohaghdam, A. A novel method for preparing aluminum diffusion coating by nanocrystalline plasma electrolysis. *Electrochemistry Communications*, 2007, 9(11), 2686-2691.

[9] Aliofkhazraei, M. and Sabour Rouhaghdam, A. Fabrication of TiC/WC ultra hard nanocomposite layers by plasma electrolysis and study of its characteristics. *Surface and Coatings Technology*, 2010.

[10] Aliofkhazraei, M., Sabour Rouhaghdam, A. and Heydarzadeh, A. Strong relation between corrosion resistance and nanostructure of compound layer of treated 316 austenitic stainless steel. *Materials Characterization*, 2009, 60(2), 83-89.

[11] Aliofkhazraei, M., Sabour Rouhaghdam, A., Heydarzadeh, A. and Elmkhah, H. Nanostructured layer formed on CP-Ti by plasma electrolysis (effect of voltage and duty cycle of cathodic/anodic direction). *Materials Chemistry and Physics*, 2009, 113(2-3), 607-612.

[12] Aliofkhazraei, M., Sabour Rouhaghdam, A. and Shahrabi, T. Abrasive wear behaviour of Si_3N_4/TiO_2 nanocomposite coatings fabricated by plasma electrolytic oxidation. *Surface and Coatings Technology*, 2010.

[13] Aliov, M.K. and Sabur, A.R. Formation of a novel hard binary SiO_2/quantum dot nanocomposite with predictable electrical conductivity. *Modern Physics Letters* B, 2010, 24(1), 89-96.

[14] Bardon, J., Bour, J., Frari, D.D., Arnoult, C. and Ruch, D. Dispersion of cerium-based nanoparticles in an organosilicon plasma polymerized coating: Effect on corrosion protection. *Plasma Processes and Polymers*, 2009, 6(SUPPL. 1), S655-S659.

[15] Barshilia, H.C., Prakash, M.S., Poojari, A. and Rajam, K.S. Corrosion behaviour of TiN/a-C superhard nanocomposite coatings prepared by a reactive DC magnetron sputtering process. *Transactions of the Institute of Metal Finishing,* 2004, 82(3-4), 123-128.

[16] Boshkov, N., Tsvetkova, N., Petrov, P., Koleva, D., Petrov, K., Avdeev, G., Tsvetanov, C., Raichevsky, G. and Raicheff, R. Corrosion behavior and protective ability of Zn and Zn-Co electrodeposits with embedded polymeric nanoparticles. *Applied Surface Science*, 2008, 254(17), 5618-5625.

[17] Castagno, K.R.L., Dalmoro, V., Mauler, R.S. and Azambuja, D.S. Characterization and corrosion protection properties of polypyrrole / montmorillonite electropolymerized onto aluminium alloy 1100. *Journal of Polymer Research*, 2009, 1-9.

[18] Chang, K.C., Chen, S.T., Lin, H.F., Lin, C.Y., Huang, H.H., Yeh, J.M. and Yu, Y.H. Effect of clay on the corrosion protection efficiency of PMMA/Na^+-MMT clay nanocomposite coatings evaluated by electrochemical measurements. *European Polymer Journal*, 2008, 44(1), 13-23.

[19] Chang, K.C., Huang, H.H., La, M.C., Hung, C.B., Chand, B., Yeh, J.M. and Yu, Y.H. Comparative electrochemical studies at different operational temperatures for the effect of nanoclay platelets on the anticorrosion efficiency of organo-soluble polyimide/clay nanocomposite coatings. *Journal of Nanoscience and Nanotechnology*, 2009, 9(5), 3125-3133.

[20] Chang, K.C., Jang, G.W., Peng, C.W., Lin, C.Y., Shieh, J.C., Yeh, J.M., Yang, J.C. and Li, W.T. Comparatively electrochemical studies at different operational temperatures for the effect of nanoclay platelets on the anticorrosion efficiency of DBSA-doped polyaniline/Na^+-MMT clay nanocomposite coatings. *Electrochimica Acta,* 2007, 52(16), 5191-5200.

[21] Chang, K.C., Lai, M.C., Peng, C.W., Chen, Y.T., Yeh, J.M., Lin, C.L. and Yang, J.C. Comparative studies on the corrosion protection effect of DBSA-doped polyaniline prepared from in situ emulsion polymerization in the presence of hydrophilic Na^+-MMT and organophilic organo-MMT clay platelets. *Electrochimica Acta*, 2006, 51(26), 5645-5653.

[22] Chaudhari, S., Patil, P.P., Mandale, A.B., Patil, K.R. and Sainkar, S.R. Use of poly(o-toluidine)/ZrO_2 nanocomposite coatings for the corrosion protection of mild steel. *Journal of Applied Polymer Science*, 2007, 106(1), 220-229.

[23] Ciubotariu, A., Benea, L., Lakatos-Varsanyi, M. and Dragan, V. Electrochemical impedance spectroscopy and corrosion behaviour of Al_2O_3-Ni nano composite coatings. *Electrochimica Acta*, 2008, 53(13), 4557-4563.

[24] Feng, Q., Li, T., Teng, H., Zhang, X., Zhang, Y., Liu, C. and Jin, J. Investigation on the corrosion and oxidation resistance of Ni-Al_2O_3 nano-composite coatings prepared by sediment co-deposition. *Surface and Coatings Technology*, 2008, 202(17), 4137-4144.

[25] Grigoriev, D.O., Kuhler, K., Skorb, E., Shchukin, D.G. and Muhwald, H. Polyelectrolyte complexes as a "smart" depot for self-healing anticorrosion coatings. *Soft Matter*, 2009, 5(7), 1426-1432.

[26] Hosseini, M.G., Raghibi-Boroujeni, M., Ahadzadeh, I., Najjar, R. and Seyed Dorraji, M.S. Effect of polypyrrole-montmorillonite nanocomposites powder addition on corrosion performance of epoxy coatings on Al 5000. *Progress in Organic Coatings*, 2009, 66(3), 321-327.

[27] Jing, C. and Hou, J. Sol-gel-derived alumina/polyvinylpyrrolidone hybrid nanocomposite film on metal for corrosion resistance. *Journal of Applied Polymer Science,* 2007, 105(2), 697-705.

[28] Kamalan Kirubaharan, A.M., Selvaraj, M., Maruthan, K. and Jeyakumar, D. Synthesis and characterization of nanosized titanium dioxide and silicon dioxide for corrosion resistance applications. *Journal of Coatings Technology Research,* 2009, 1-8.

[29] Khazrayie, M.A. and Aghdam, A.R.S. Si_3N_4/Ni nanocomposite formed by electroplating: Effect of average size of nanoparticulates. *Transactions of Nonferrous Metals Society of China*, 2010, 20(6), 1017-1023.

[30] Khazrayie, M.A. and Aghdam, A.S.R. Characterization of Ni-W/MWCNT nanocomposite layers formed by pulsed electrochemical deposition. *Protection of Metals*, 2010(6).

[31] Kozhukharov, V., Kozhukharov, S., Tsaneva, G., Gerwann, J., Schem, M., Schmidt, T. and Veith, M. Investigation on the corrosion protection ability of nanocomposite hybrid coatings. *Bulgarian Chemical Communications,* 2008, 40(3), 310-317.

[32] Li, J., Sun, Y., Sun, X. and Qiao, J. Mechanical and corrosion-resistance performance of electrodeposited titania - Nickel nanocomposite coatings. *Surface and Coatings Technology*, 2005, 192(2-3), 331-335.

[33] Ma, D., Wang, X., Ma, S. and Xu, K. Electrochemical corrosion of Ti-Si-N films coated by pulsed-DC plasma enhanced CVD. Xiyou Jinshu Cailiao Yu Gongcheng/*Rare Metal Materials and Engineering*, 2004, 33(7), 740-743.

[34] Ma, J., Shi, Y., Di, J., Yao, Z. and Liu, H. Effect of TiO_2 nanoparticles on anticorrosion property in amorphous Ni-P-Cr composite coating in artificial seawater and microbial environment. *Materials and Corrosion*, 2009, 60(4), 274-279.

[35] Malucelli, G., Di Gianni, A., Deflorian, F., Fedel, M. and Bongiovanni, R. Preparation of ultraviolet-cured nanocomposite coatings for protecting against corrosion of metal substrates. *Corrosion Science*, 2009, 51(8), 1762-1771.

[36] Mirzamohammadi, S., Aliov, M.K., Sabur, A.R. and Hassanzadeh-Tabrizi, A. Study Of Wear Resistance And Nanostructure For Tertiary Al_2O_3/Y_2O_3/CNT Pulsed Electrodeposited Ni-Based Nanocomposite. *Materials Science*, 2010(1).

[37] Mirzamohammadi, S., Kiarasi, R., Aliov, M.K., Sabur, A.R. and Hassanzadeh-Tabrizi, A. Study of corrosion resistance and nanostructure for tertiary Al_2O_3/Y_2O_3/CNT pulsed electrodeposited Ni based nanocomposite. *Transactions of the Institute of Metal Finishing*, 2010, 88(2), 93-99.

[38] Nguyen, V.N., Perrin, F.X. and Vernet, J.L. Water permeability of organic/inorganic hybrid coatings prepared by sol-gel method: A comparison between gravimetric and capacitance measurements and evaluation of non-Fickian sorption models. *Corrosion Science*, 2005, 47(2), 397-412.

[39] Park, J.H., Kwon, S.H., Lee, M.H. and Kim, K.H. Effect of Si Addition on the Corrosion Behavior of Ti-Si-N Coatings Prepared by a Hybrid Coating System. *Electrochemical and Solid-State Letters*, 2009, 12(3), C13-C15.

[40] Pirhady Tavandashti, N., Sanjabi, S. and Shahrabi, T. Corrosion protection evaluation of silica/epoxy hybrid nanocomposite coatings to AA2024. *Progress in Organic Coatings*, 2009, 65(2), 182-186.

[41] Polychronopoulou, K., Baker, M.A., Rebholz, C., Neidhardt, J., O'Sullivan, M., Reiter, A.E., Kanakis, K., Leyland, A., Matthews, A. and Mitterer, C. The nanostructure, wear and corrosion performance of arc-evaporated CrB_xN_y nanocomposite coatings. *Surface and Coatings Technology*, 2009, 204(3), 246-255.

[42] Praveen, B.M., Venkatesha, T.V., Naik, Y.A. and Prashantha, K. Corrosion behavior of Zn-TiO_2 composite coating. *Synthesis and Reactivity in Inorganic, Metal-Organic and Nano-Metal Chemistry*, 2007, 37(6), 461-465.

[43] Qiang, M., Chen, L., Chen, T. and Huang, F. Corrosion resistance of polyaniline / monmorillonite nanocomposite material. *Corrosion and Protection,* 2005, 26(5), 203-204+211.

[44] Raj, V. and Mubarak Ali, M. Formation of ceramic alumina nanocomposite coatings on aluminium for enhanced corrosion resistance. *Journal of Materials Processing Technology*, 2009, 209(12-13), 5341-5352.

[45] Rosero-Navarro, N.C., Pellice, S.A., Castro, Y., Aparicio, M. and Durán, A. Improved corrosion resistance of AA2024 alloys through hybrid organic-inorganic sol-gel coatings produced from sols with controlled polymerisation. *Surface and Coatings Technology*, 2009, 203(13), 1897-1903.

[46] Shi, L., Sun, C., Gao, P., Zhou, F. and Liu, W. Mechanical properties and wear and corrosion resistance of electrodeposited Ni-Co/SiC nanocomposite coating. *Applied Surface Science*, 2006, 252(10), 3591-3599.

[47] Shi, L., Zhou, F., Sun, C.F. and Liu, W.M. Corrosion resistance and tribological properties of NiCo-SiC nanocomposite coatings. Zhongguo Youse Jinshu Xuebao/*Chinese Journal of Nonferrous Metals*, 2005, 15(4), 536-540.

[48] Sugama, T. Polyphenylenesulfied/montomorillonite clay nanocomposite coatings: Their efficacy in protecting steel against corrosion. *Materials Letters*, 2006, 60(21-22), 2700-2706.

[49] Szeptycka, B. and Gajewska-Midzialek, A. The influence of the structure of the nanocomposite Ni-PTFE coatings on the corrosion properties. *Reviews on Advanced Materials Science*, 2007, 14(2), 135-140.

[50] Tallman, D.E., Levine, K.L., Siripirom, C., Gelling, V.G., Bierwagen, G.P. and Croll, S.G. Nanocomposite of polypyrrole and alumina nanoparticles as a coating filler for the corrosion protection of aluminium alloy 2024-T3. *Applied Surface Science*, 2008, 254(17), 5452-5459.

[51] Tian, W., Wang, Y., Zhang, T. and Yang, Y. Sliding wear and electrochemical corrosion behavior of plasma sprayed nanocomposite Al_2O_3-13%TiO_2 coatings. *Materials Chemistry and Physics*, 2009, 118(1), 37-45.

[52] Vaezi, M.R., Sadrnezhaad, S.K. and Nikzad, L. Electrodeposition of Ni-SiC nano-composite coatings and evaluation of wear and corrosion resistance and electroplating characteristics. *Colloids and Surfaces A: Physicochemical and Engineering Aspects,* 2008, 315(1-3), 176-182.

[53] Wang, T.L., Hwang, W.S. and Yeh, M.H. Preparation, properties, and anticorrosion application of poly(methyl methacrylate)/montmorillonite nanocomposites coating on

brass via solution polymerization. *Journal of Applied Polymer Science,* 2007, 104(6), 4135-4143.

[54] Wang, W., Qian, S.Q., Zhou, X.Y., Lin, W.S. and Yan, M.J. Microstructure and corrosion properties of Ni/PTFE nanocomposite coatings by high speed jet electrodeposition. Jinshu Rechuli/*Heat Treatment of Metals*, 2006, 31(SUPPL.), 149-152.

[55] Wang, Y., Lim, S., Luo, J.L. and Xu, Z.H. Tribological and corrosion behaviors of Al_2O_3/polymer nanocomposite coatings. *Wear*, 2006, 260(9-10), 976-983.

[56] Yang, L., Liu, F. and Han, E. Anti-corrosion performance studies of the nano-ZnO polyurethane coatings. Cailiao Yanjiu Xuebao/*Chinese Journal of Materials Research,* 2006, 20(4), 354-360.

[57] Yao, Y. Effect of SiO_2 nano-particulates on corrosion behaviour of Ni-W/SiO2 nanocomposite coatings. *Anti-Corrosion Methods and Materials,* 2007, 54(6), 336-340.

[58] Yao, Y., Yao, S. and Zhang, L. Corrosion behavior of Ni-W/SiC nanocomposite coating in NaCl solution. *Surface Review and Letters*, 2006, 13(4), 489-494.

[59] Yao, Y., Yao, S., Zhang, L. and Wang, H. Electrodeposition and mechanical and corrosion resistance properties of Ni-W/SiC nanocomposite coatings. *Materials Letters*, 2007, 61(1), 67-70.

[60] Yeh, J.M., Chen, C.L., Chen, Y.C., Ma, C.Y., Huang, H.Y. and Yu, Y.H. Enhanced Corrosion Prevention Effect of Polysulfone-Clay Nanocomposite Materials Prepared by Solution Dispersion. *Journal of Applied Polymer Science,* 2004, 92(1), 631-637.

[61] Yeh, J.M., Kuo, T.H., Huang, H.J., Chang, K.C., Chang, M.Y. and Yang, J.C. Preparation and characterization of poly(o-methoxyaniline)/Na^+-MMT clay nanocomposite via emulsion polymerization: Electrochemical studies of corrosion protection. *European Polymer Journal*, 2007, 43(5), 1624-1634.

[62] Yeh, J.M., Liou, S.J., Lu, H.J. and Huang, H.Y. Enhancement of Corrosion Protection Effect of Poly(styrene-co- acrylonitrile) by the Incorporation of Nanolayers of Montmorillonite Clay into Copolymer Matrix. *Journal of Applied Polymer Science*, 2004, 92(4), 2269-2277.

[63] Yeh, J.M., Yao, C.T., Hsieh, C.F., Lin, L.H., Chen, P.L., Wu, J.C., Yang, H.C. and Wu, C.P. Preparation, characterization and electrochemical corrosion studies on environmentally friendly waterborne polyurethane/Na^+-MMT clay nanocomposite coatings. *European Polymer Journal*, 2008, 44(10), 3046-3056.

[64] Yu, Q., Ma, X., Wang, M., Yu, C. and Bai, T. Influence of embedded particles on microstructure, corrosion resistance and thermal conductivity of CuO/SiO_2 and NiO/SiO_2 nanocomposite coatings. *Applied Surface Science*, 2008, 254(16), 5089-5094.

[65] Yu, Y.H., Yeh, J.M., Liou, S.J., Chen, C.L., Liaw, D.J. and Lu, H.Y. Preparation and properties of polyimide-clay nanocomposite materials for anticorrosion application. *Journal of Applied Polymer Science*, 2004, 92(6), 3573-3582.

[66] Zaarei, D., Sarabi, A.A., Sharif, F. and Kassiriha, S.M. Structure, properties and corrosion resistivity of polymeric nanocomposite coatings based on layered silicates. *Journal of Coatings Technology Research*, 2008, 5(2), 241-249.

[67] Zamblau, I., Varvara, S., Bulea, C. and Muresana, L.M. Corrosion Behavior of CompositeCoatingsObtained by ElectrolyticCodeposition of Copperwith Al_2O_3 Nanoparticles. *Chemical and Biochemical Engineering Quarterly,* 2009, 23(1), 43-52.

[68] Zandi-Zand, R., Ershad-Langroudi, A. and Rahimi, A. Silica based organic-inorganic hybrid nanocomposite coatings for corrosion protection. *Progress in Organic Coatings*, 2005, 53(4), 286-291.

[69] Zhang, W., Li, L., Yao, S. and Zheng, G. Corrosive mechanism of paint coatings modified by carbon black nanoparticles in NaCl solution. *Journal of the Chinese Society of Corrosion and Protection*, 2005, 25(6), 351-355.

[70] Zhang, W.G., Li, L., Yao, S.W. and Zheng, G.Q. Evaluation of corrosion resistance of nanometer carbon black composite coating by electrochemical method. Cailiao Gongcheng/*Journal of Materials Engineering*, 2006(8), 49-51.

[71] Zhang, Y., Zhang, C., Peng, X. and Wang, F. Effect of Cr particle size on the oxidation behaviour of electrodeposited Ni-Cr composite coatings. *Journal of the Chinese Society of Corrosion and Protection*, 2006, 26(2), 85-88.

[72] Zhou, Y.B., Peng, X. and Wang, F.H. Oxidation of a novel electrodeposited Ni - 28.0 mass% Al nanocoating. *Corrosion Science and Protection Technology*, 2005, 17(4), 219-222.

Chapter 6

MECHANICAL PROPERTIES OF NANOCOMPOSITE COATINGS

ABSTRACT

Mechanical properties of nanocomposite coatings such as hardness, roughness, wear resistance and some examples were discussed in this chapter. It was focused on both metallic and ceramic matrix composites in the examples. There are some various techniques for producing nanocomposite coatings. Electrodeposition is a method for simultaneous deposition of particle in micron size or smaller than micron which made of metal, non-metallic or polymeric compounds along metallic or alloyed matrix that can creates thick deposits with extended surfaces up to thin deposits in nano dimensions. In recent decades, this works aimed at producing resistant coatings against erosion and corrosion in self-lubricating systems and hardness diffused coatings. Electrodeposition, as the surface final process can be replaced by other surface processes like hard chromium.

6.1. INTRODUCTION

Industrial work-pieces are always exposed to corrosive environments and mechanical complex, thermal, and erosive forces etc. which particularly appear on their surfaces. Base material could not often bear such forces and environments since economic concerns do not allow using a great amount of the resistant materials and it is only preferred dealing with the surface and/or type of application does not permit to use resistant material so one should think about solution for prevention from structure degradation. One of such solution is to use appropriate coating on pieces surface. With respect to application type, appropriate coatings will be used. One of such coatings is metal matrix composite coatings comprising diffused particles. Metal Matrix Composite (MMC) is a kind of materials in which the metal-based properties are improved by adding different materials. Composite coatings have superior properties in terms of hardness, resistance against erosion, oxidation and corrosion and self-lubricating which have allowed them to be used in different applications.

There are some various techniques for producing such coatings where one of important methods is electrodeposition. Electrodeposition is a method for simultaneous deposition of particle in micron size or smaller than micron which made of metal, non-metallic or polymeric compounds along metallic or alloyed matrix that can creates thick deposits with extended surfaces up to thin deposits in nano dimensions. In recent decades, this works aimed at producing resistant coatings against erosion and corrosion in self-lubricating systems and hardness diffused coatings. Electrodeposition, as the surface final process can be replaced by other surface processes like hard chromium.

6.2. History of Metal Matrix Composite Coatings

Many researchers have succeeded to produce different simultaneous deposits on some metals like Ni, Cr, Co etc. These particles may include hard oxides like Al_2O_3, TiO_2, SiO_2, CeO_2, some carbides like SiC, WC, diamond, and some nitrides such as TiN and solid lubricants like PTFE and MoS_2. In fact, this type of coating was known since nineteenth century. In 1928, graphite-copper layers were tested to use in car engine. Since the late 1950's, nickel coating comprising of diffused particles was produced, even though its process theory had not been yet known. One could study metallic coatings with internal particles following the related researches about copper-graphite composite at self-lubricating level. In 1950 and 1960's, production of composite coatings was developed.

During two recent decades, composite coating has advanced remarkably. Its first application is related to using Ni/SiC coating for improvement of resistance against erosion and since that time Ni/SiC composite coating has been successfully used in automotive industries. In 1970 and 1980's, these studies were focused on production of coatings with excellent mechanical, corrosive and erosive properties. In 1990's, the advance physical properties of composite coatings were promptly characterized and new fields like electro-catalyst and photo electro-catalyst have been addressed. During recent years, due to their superior properties like thermodynamic stability, electrical conductivity, mechanical, physical, magnetic properties etc, study on nano-structural materials has been increased. By further variation of particle in nano dimensions and in parallel to using nanotechnology science for developing thin coatings, application of nano-particles was also prevalent in composite coatings. The presence of nano-particles on the border of grains may impede movement of dislocations and recrystallization at higher temperature. Therefore, micro-hardness and thermal stability may be noticeably increased. Due to improvement in crystalline structure and increasing hardness, the resistance against erosion also promotes. Nevertheless, producing of replicable nano-composites is still difficult. On the other hand, hydrodynamic conditions may affect on particles quantity and propagation, and also due to agglomeration of particles, it is necessary to use ultrasonic conditions.

6.3. The Aim of Composite Coating Production

To date, this idea has been considered that hardened diffused nickel coatings can be replaced by chromium coatings. Most of the researches around this kind of coatings is based on their hardness rather than their corrosion resistance. Corrosion resistance in these products may be due to fineness of the grains. At the atmospheres with high amount of oxygen, Ni-Al_2O_3 composite coatings have better oxidation resistance than disarmed nickel. In application of Ni-SiC the improvement in erosive resistance has been reported. Al_2O_3 and SiO_2 oxide thin particles may be electrodeposited along nickel by watts and sulfamate baths. One could assume the improvement in hardness, resistance against erosion and corrosion as one of the most important goals for fabrication of composite coatings.

6.3.1. Hardness Increase

The hardness of commercial pure nickel deposit is about 215 to 265 Vickers while composite coating hardness varies from 275 to 850 Vickers. In fact, the presence of the particles in this coating may increase hardness at this coating. It has been characterized that composite coating hardness and other mechanical properties not only depend on mechanical properties of matrix and particles, but also on quantity and size of particles. Quantity and size of particles in both mechanisms may determine reinforcement in metallic matrix composites. It was found that coating hardness may increase by increase in the quantity of coating particles. This result was given in figure 6.1. Also in figure 6.2, coating hardness increase due to increasing quantity of silicon carbide has been illustrated.

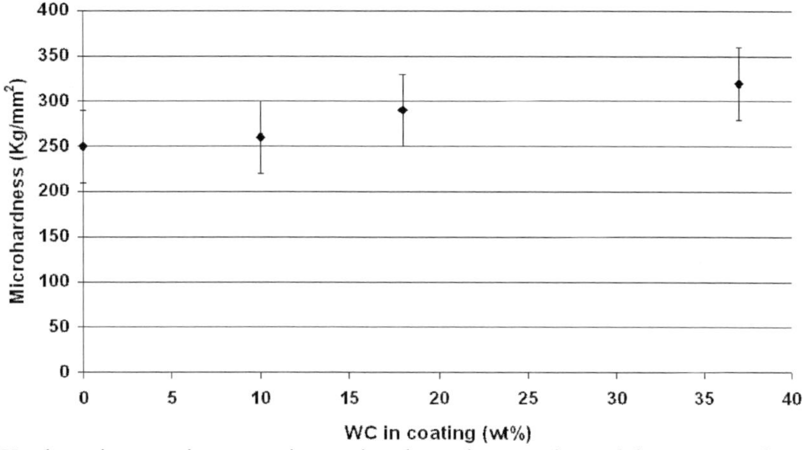

Figure 6.1. Hardness increase in composite coating due to increase in particles amount of WC.

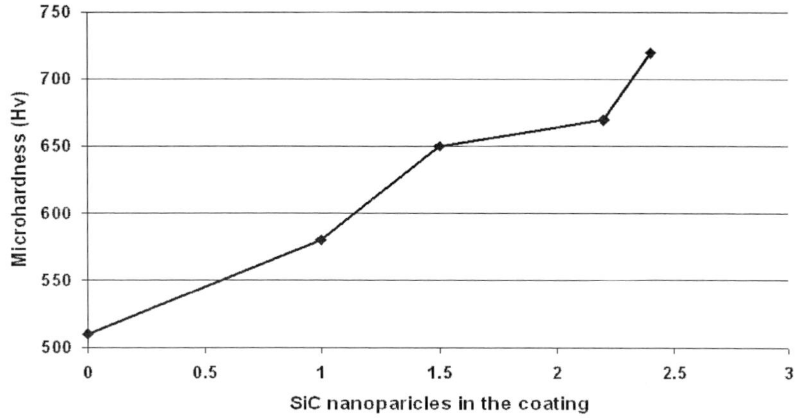

Figure 6.2. Deposit hardness increase due to quantity increase of silicon carbide.

One could express different mechanisms for the way of increase in deposits hardness due to presence of particles.

6.3.1.1. Strengthening Mechanisms in Nano-Composites

In general condition, the strengthening mechanism for metallic or alloyed poly-crystals is as follows:

I. Strengthening due to grains fineness according to Hall-Petch formula;
II. Diffusion hardness according to Orowan mechanism;
III. Strengthening of solid solution; and
IV. Crystallographic orientation

Solid solution is seen in some cases like alloyed coatings. Regarding to crystallographic orientation, one could imply that it has been observed in some cases the pure nickel samples with identical grain size show various quantities of micro-hardness. This difference in hardness is originated from crystalline directions. The harder samples have crystalline directions (111), while the crystalline orientation is (200) in softer samples. About composite coatings, because of observing small peaks during exploration by X-ray, one could ignore the orientation effect as well, and one could assume coating hardness due to grains fineness plus hardness diffusion effect.

Hall- Petch Mechanism

By particles fineness, creation and movement of dislocations are limited so this increases the strength. In this model, grains boundary acts as a barrier against moving dislocations so this leads to dislocations pile-up behind grain boundary. In many nano-crystalline materials, high hardness has been reported in relation to Hall-Petch effect. One could write strengthening effect due to grains fineness as follows:

$$HV = 3\sigma_\gamma$$

$$\sigma_Y = \sigma_0 + kd^{-1/2}$$

where σ_Y denotes yield tension, σ_0 is frictional tension, *k* is constant and d is grains diameter.

Concerning to some materials an unexpected accident occurs and by being finer than certain rate, the material will be softer (inverse Hall-Petch behavior). In this case, it was reported that due to reduction of grains size from 20 to 3 nm, Ni-P alloy was softened and the maximum hardness which was reported for granular size was 8 nm. By reduction in grain size from 100 to 2 nm, the expansion of grains boundary increases. Similarly, through increase in the size of grains triple boundaries, deflection is created in Hall-Petch models and these boundaries affect more than grains boundary expansion.

Orowan Mechanism

One could write Orowan diffusion hardness as follows:

$$\tau = 2Gb/\lambda$$

In this formula, G and b denote grid shear modulus and dislocation berger's vector. λ refers to distance between particles. This equation shows that particles should be diffused well in order to maximize strength.

6.3.2. Increasing Resistance against Erosion

Erosive resistance and hardness are interdependent and increase in hardness leads to increase in erosive resistance. As coating hardness is increased by increasing particles quantity in deposit, it also increases erosive resistance in particles. This means the weight of removed coating is reduced due to erosion. The given result is illustrated in figure 6.3. The more particles are placed in the deposit, the lesser grooves will be created by erosion.

The presence of such grooves may lead to increasing erosion. Due to sinking of erosive particles residue over the deposit, some cracks are created in parallel to erosion direction along erosive grooves. Due to erosion of composite coating surface, the particles are exited from coating. By exploration into erosive effects on composite coatings, it is concluded that if these particle remain through erosive surfaces, the given particles affect negatively on erosive resistance; in fact, there is a competition among hardness increase and surface erosion. By increase in quantity of particles in deposit, hardness increases further, but despite of this fact if the particles are ejected, the surface is eroded more. The presence of hard particles in coating causes reduced friction among erosive object and nickel soft grid. In figure 6.4, the effect of tungsten carbide on reduced friction coefficient was illustrated. This reduction in friction coefficient may also increase erosive resistance.

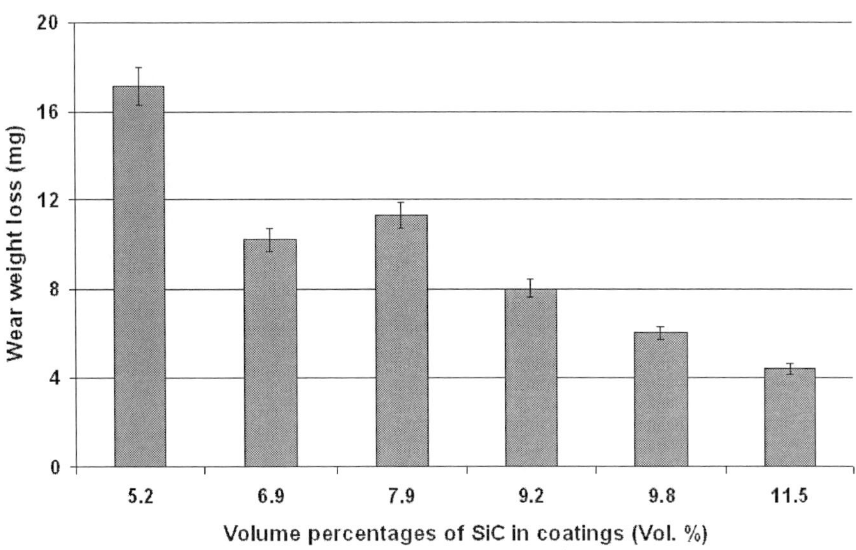

Figure 6.3. Increasing erosive resistance due to increase in quantity of deposited particles.

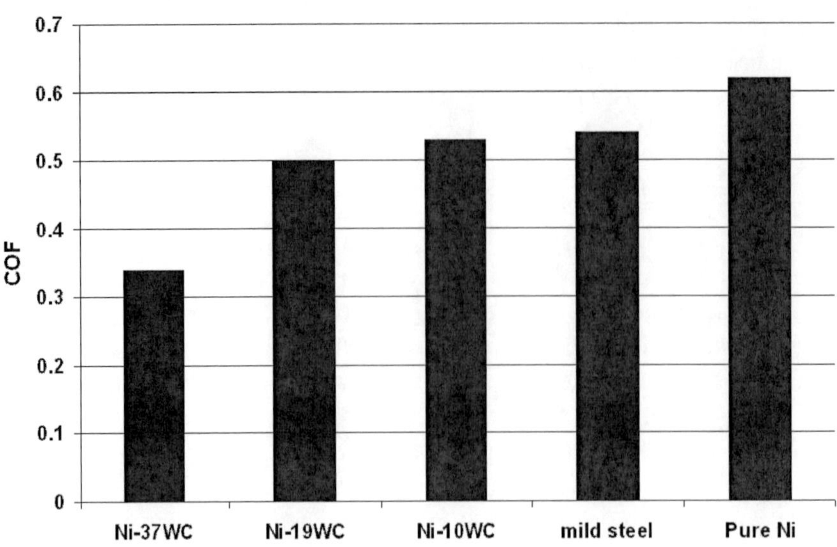

Figure 6.4. Reduced friction coefficient due to increase in WC in Ni-WC composite coating.

6.3.3. Increase in Resistance against Corrosion

When working in corrosive environments, corrosion resistance of coating like hardness and wear resistance and even more than both is also very important. Corrosion resistance in composite coating in different conditions like dipping (immersion) in common salt solution 3.5% and/or with the presence of erosive condition have been examined. Secondary particles which are initially placed in electrodeposition bath and then in coating are responsible for creation of secondary properties in coating that affect on corrosion resistance as well. They change coating situation like surface smoothness, porosity and density. They generate tension and lead to creation of some faults while changing coating crystalline structure. Type and quantity of adjoined particles, the existing impurity in them and particles compatibility include some items which affect on coating corrosive properties. Particle uniform propagation within metallic lattice is also important. By interrupting nickel crystals regular growth and creation of nucleation new centers, these particles affect on nickel crystallization. Such distortion in matrix structure may lead to create some cracks, bores and internal tension which cause reduced corrosion resistance while increasing sectional corrosion event and tensional corrosion. On the other hand, this causes to intensified oxygen presence and accelerated nickel passivity. Existing impurities in particles or in bath may also speed up nickel passivity. The fact that adjoined particles may affect on surface morphology and matrix protective properties through different methods has led to present different reports and often contradictory ones about corrosion resistance in composite coatings.

In many environments, nickel coatings show a good corrosion resistance quality; the environments like atmosphere, basic, molten hydroxides, non-oxidative diluted mineral oxides and higher temperatures (higher than 875 °C). Nickel corrosion resistance is due to formation of oxide layer on the surface. Chloride ions may destroy this layer so sectional cells are formed on the surface. The latest researches indicate the better corrosion resistance in Ni-SiC and Ni-Al_2O_3 than pure nickel in salt water environment (salt water 0.3 and 0.6 molar). It is observed that dielectric particles of Al_2O_3 may accelerate formation of passive layer in nickel matrix and if these particles are finer because of non- conductivity they act as barriers if they are separated homogeneously in nickel matrix and they logically reduce the link between matrix and corrosive environment; therefore, the finer particles cause better corrosion resistance in composite. If the particles are conductive, their corrosive behavior depends on their electrochemical potential. During exploration on corrosion resistance of titanium oxide-nickel in common salt solution 3.5% it was characterized that because of the presence of these particles in the coating, corrosive potential approaches to more positive quantities which means nobler behavior in coating. Similarly, due to increase in corrosion resistance, corrosion current density reduces. Investigation into electrochemical impedance also confirms this result and the electrochemical impedance for TiO_2-Zn coating in common salt (3.5%) is also confirmed this conclusion. Also, these coatings have greater resistance (R_P) and lesser capacity of dual layer (C_{dl}) that confirms their higher corrosion resistance. These results were confirmed by other researchers and concerning to other composite coatings.

6.4. The Aim of Ni-WC Nano-Composite Coating Fabrication on Copper

Due to its extraordinary electrical and thermal conductivity, copper has several applications but its wear resistance is low so it should be improved by some methods. Hard chromium corrosion resistant coatings are widely used but because of existing chromium ions (Cr^{6+},Cr^{3+}) in electrodeposition electrolyte, one should find alternative for these coatings. The adjoined project by US Ministry of Defense and Canadian Ministry of Defense has been carried out where instead of hard chromium, High Velocity Oxygen Fuel (HVOF) (WC Spray) or WC constituent coatings were used in aircraft landing gear wheels. Tungsten Carbide (WC) or WC-Co is a technological material which is widely used in cutting tools, drill, punch and hard coating materials. Ni-WC coating is an appropriate option since carbide hardness is incorporated with matrix toughness.

There are some different methods to perform WC coating; these techniques include:

1. Hard facing;
2. Vapor deposition (in vacuum);
3. High Velocity Oxygen Fuel (HVOF) spray; and
4. Electrodeposition

- *Hard facing:* This is a normal technique for WC-based coating. Unfortunately, it is hard for producing thin and continuous coatings. Also objects shape is another restrictive factor in this method. Similarly, it is difficult task to achieve nano-structure by this technique, since there is no mechanism for stopping crystal growth as well as it is not possible to control metallurgical changes during this process.
- *Vapor deposition (in vacuum):* It is conducted by two physical (PVD) and chemical (CVD) methods. It is also very expensive.

- *Spray (HVOF):* It is a various technique for creation of coating which is used to generate different alloys coating in many sensitive parts of refinery, mine etc. industries. By this method it may produce coating with more strengthened and less porous links than other spray methods. The great limitation of this method is its higher temperature which approaches to 2538 °C. Such high temperature may easily melt refractory powders like WC and cause unfavorable metallurgical changes (like phase change). Although high under-cooling between mid-flame and substrate creates a production matrix with a very fine structure but it is inevitable to create severe thermal tensions.
  The common limitation of all three methods is inaccessibility to internal small facings, internal edges and extraordinary disorderly forms. For instance in spray method, it is not possible to access to pieces corner or edge and internal surfaces of pipes with small diameter. To produce coating with nano-structure, nano-structural powders with spraying capacity should exist. Also, powders with low weight which causes difficulty in their conveyance by gas and achieving to this goal as well as existing problems in control of chemical and structural changes during the process converted spray method into a non-preferred technique. On the other hand, the created coatings by spray which are widely used on steel are not performable because of copper high thermal conductivity.
- *Electrodeposition:* This method provides an atomic bound between coating and matrix which is stronger than mechanical link which created by spray method. Electrodeposition is done easily and executable for all geometrical figures. By arising of nano-structure materials during three last decades, electrochemical deposition techniques provided a certain path for manufacture of different kinds of nano-structured materials. These varieties comprised of nano-crystalline deposits, nano-wires, nano-tubes, nano-layers and nano-composites.

Basically, electrochemical deposition leads to production of nano-structures. If the parameters in this process (like bath composition, pH, temperature, over-potential etc.) are selected in such a way that electrochemical crystallization provides fast nucleation and low growth, the electrochemical crystallization is carried out through two competitive processes (i.e. reconstruction of the present crystals and formation of new ones) that are affected by several factors. The speed of charge transfer and surface diffusion of adjoined ions on crystal surface are the most important determinant stages. In low over-potentials and surface diffusion high rates, the crystal growth process predominates while in high over-potentials and surface diffusion low rates, new nucleation forms. Due to having very fine grains and high nucleation rate, nano-crystalline materials have extraordinary mechanical, physical, chemical and electrochemical properties than amorphous materials or conventional poly crystals. Grain reduced size also leads to severe increase in nano-crystalline materials erosive resistance. For WC-Co nano-structure composites, for example, reduction of WC grains size to 70 nm causes acquiring wear resistance to two times more than in normal drills.

6.5. Examples of Tribological Improvement by Nanocomposite Coatings

6.5.1. Ni-W/Al_2O_3/CNT Nanocomposite Coating

Figure 6.5 illustrates the effect of different average size of nanoparticles (ASNP) on the R_a of coatings. Interpolated equation shows that there is a quadratic relation among the roughness of obtained layer and ASNP. It can be concluded that the interaction among nanoparticles with low ASNP (approximately less than 90 nm) will increase the roughness of obtained layer. The minimum roughness has been obtained for the nanocomposite layer with ASNP equal to 93 nm.

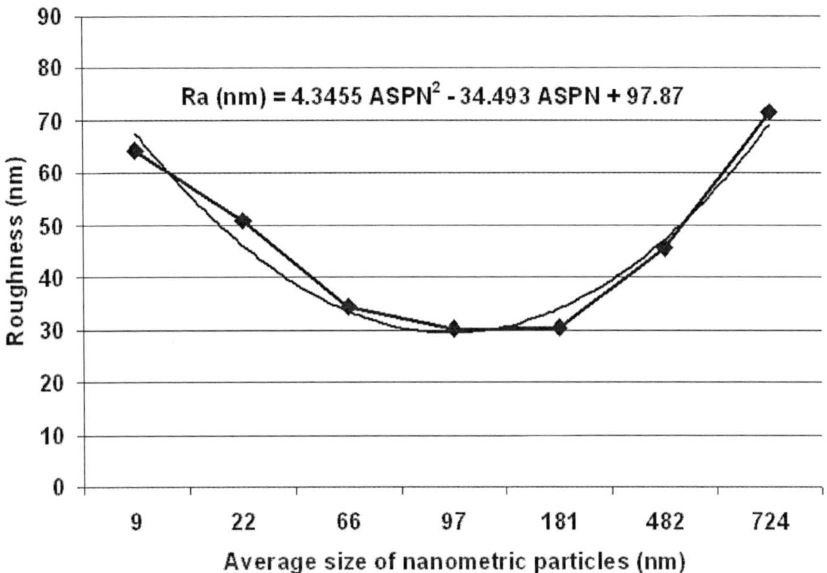

Figure 6.5. Effect of different ASNPs on the R_a of coatings.

6.5.2. Ni/Al_2O_3/Y_2O_3/CNT Nanocomposite Coating

An optimization method for the design of experiment has been used for optimizing tertiary (Al_2O_3/Y_2O_3/CNT) nanocomposite electrodeposited coating process parameters for wear protection of treated samples. The contribution of Y_2O_3 concentration is more than the sum of the contributions of all the other three factors. It is evident that, among the selected factors, Y_2O_3 concentration has the major influence on the wear rate of performed coatings. It can be seen that the current density is the second important factor that affects the wear rate of the treated substrates. Furthermore, it can be assumed that treatment time and temperature of electrolyte have almost the same effect on wear rates of coatings because of the minor difference in the contribution percentages between these two factors. By ranking their relative contributions, the sequence of the four factors affecting the wear rate is Y_2O_3 concentration, current density, treatment time and temperature of electrolyte. In the case of average size of nanoparticles ranking of effective factors by their relative contributions is the same as for wear rate which shows strong relation between these two measured properties of coatings. AFM analysis confirmed smooth surface and average size of nanoparticles in the optimal coating.

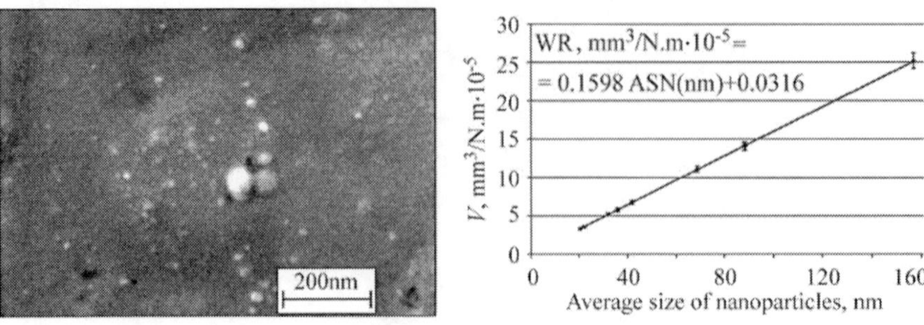

Figure 6.6. (a) SEM nanostructure of optimal coating (b) relation between average size of nanoparticles and wear rate.

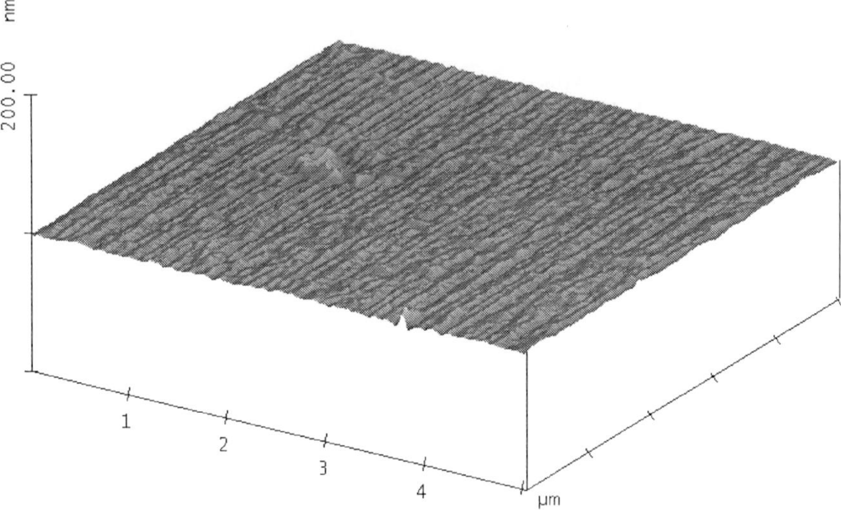

Figure 6.7. AFM nanostructure of optimal coating.

6.5.3. Ni/Si_3N_4 Nanocomposite Coating

Electrodeposited Ni/Si_3N_4 nanocomposite layers have been synthesized successfully through electrolytic deposition from a sulphate bath and the effect of average size of used nanometric particulates (ASNP) in the range of submicron scale (less than 1 μm) to nanometric scale (less than 10 nm) have been studied. The roughness illustrated a minimum level while the distribution of nanometric particulates in obtained nanocomposite layer will be more uniform by decreasing the ASNP. Response surface methodology has been proved to be accurate in prediction and optimization of the average sizes of nanoparticles. The effects of frequency, duty cycle and concentration of nanoparticulates in electrodeposition bath have been discussed and the proposed equation has been obtained by relative software.

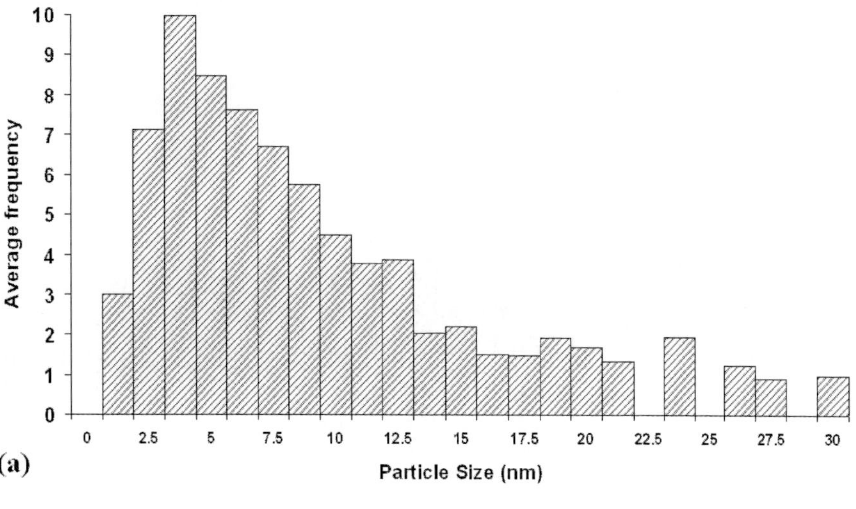
Average frequency
0
1
2
3
4
5
6
7
8
9
10
0
2.5
5
7.5
10
12.5
15
17.5
20
22.5
25
27.5
30
(a)
Particle Size (nm)

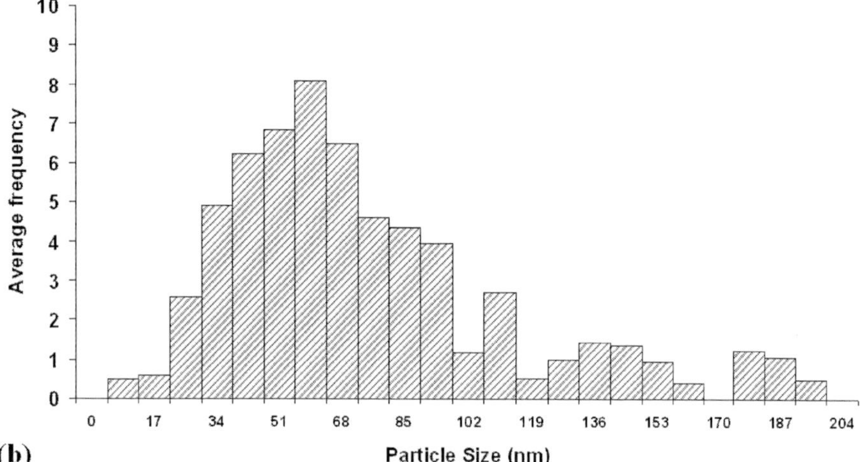
Average frequency
0
1
2
3
4
5
6
7
8
9
10
0
17
34
51
68
85
102
119
136
153
170
187
204
(b)
Particle Size (nm)

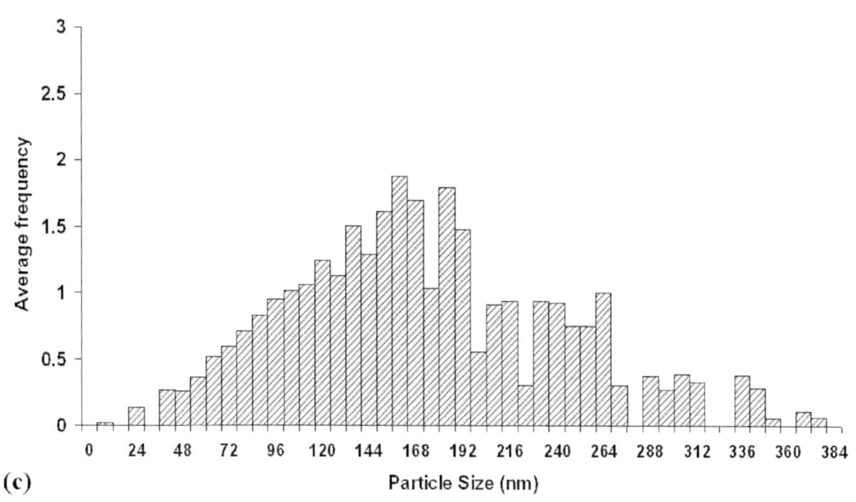
Average frequency
0
0.5
1
1.5
2
2.5
3
0
24
48
72
96
120
144
168
192
216
240
264
288
312
336
360
384
(c)
Particle Size (nm)

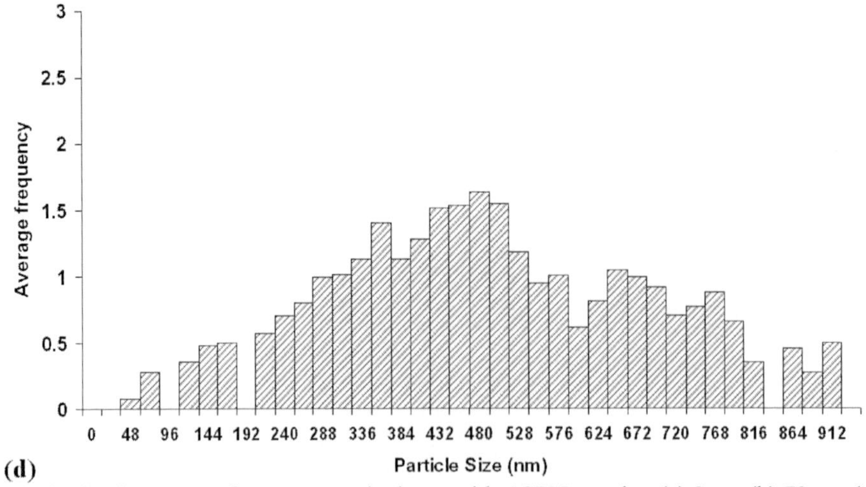

Figure 6.8. Distribution curve of nanocomposite layer with ASPN equal to (a) 9 nm (b) 72 nm (c) 168 nm (d) 499 nm.

6.5.4. TiC/WC Nanocomposite Coating

A typical surface profile of the fabricated layers for ceramic-based nanocomposite layer by plasma electrolysis can be seen in figure 6.9. This profile is relatively smooth with no sharp falling (related to the possible mini-cracks). It is also an example of different measurements, which, in addition to SEM figure, confirms a uniform surface of nanocomposite layer with no sharp cracks. The presence of nanoparticles in the fabricated layer will avoid propagation of sharp cracks. It is even useful for increasing fatigue resistance of the coatings as shown for other nanocomposite layers.

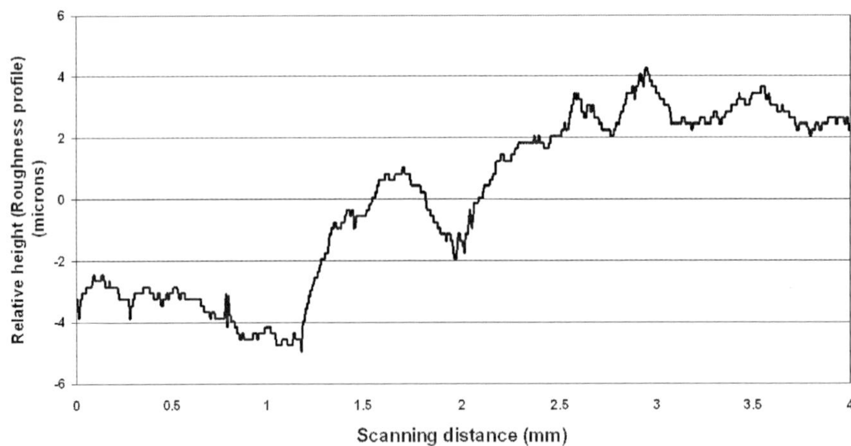

Figure 6.9. Typical surface linear profile of nanocomposite layer.

More uniform distributed nanoparticles will act better in this phenomenon. Roughness values of the nanocomposite layers were calculated to be approximately between 1.6 and 4.9 microns. Figure 6.10 illustrates the changes of roughness with the change in the concentration of WC nanoparticles in the electrolyte. Concentration of WC nanoparticles in the electrolyte has an optimum level for achieving the minimum roughness on the surface of the nanocomposite layer at higher current densities. In fact, the increase of nanoparticles concentration in the electrolyte and the increase of the current densities have similar effects on surface roughness. Higher current densities will lead to big sparks with more damaging effects and their effects will show themselves on low concentrations of nanoparticles in the electrolyte. For the layers fabricated by 2 to 6 gr/lit of WC nanoparticles in electrolyte, the increase of roughness is lesser for higher current densities as compared with high increase of roughness at lower current densities. This effect can be seen in the left side of the figure 6.10 and in different slopes of the trend lines. From this point of view, 6 gr/lit is an optimum concentration of the nanoparticles for higher current densities. Concentrations of nanoparticles in electrolyte higher than 6 gr/lit will lead to sharp increase for roughness of nanocomposite coatings.

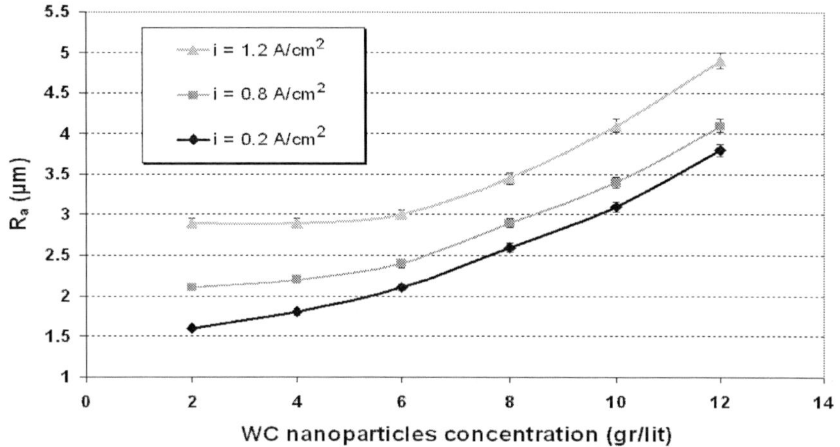

Figure 6.10. Relation among surface roughness of coating and WC nanoparticle concentration in electrolyte in different current densities.

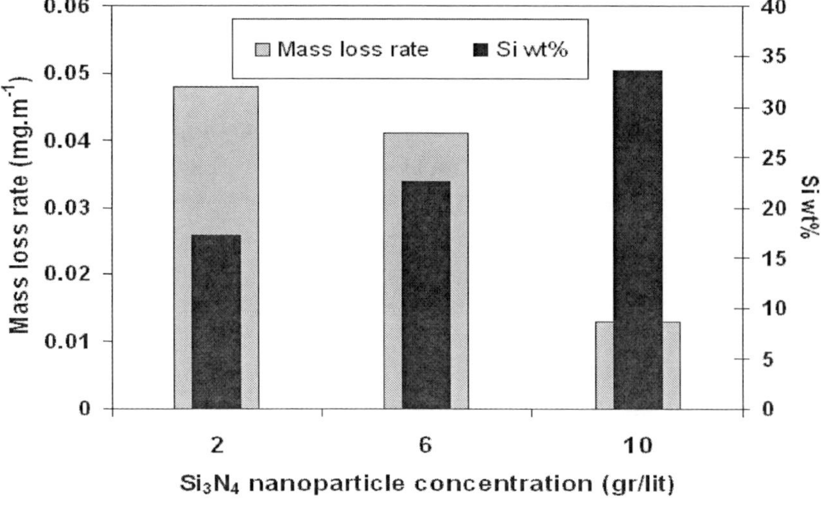

Figure 6.11. Effect of Si_3N_4 nanoparticles concentration in electrolyte on the mass loss rate and silicon content of coatings.

6.5.5. TiO_2/Si_3N_4 Nanocomposite Coating

TiO_2/Si_3N_4 nanocomposite coatings were successfully fabricated by plasma electrolytic oxidation of commercially pure titanium samples in a suspension of fine Si_3N_4 nanoparticles. Layer by layer X-ray diffractions confirmed the formation of aluminate titanate phase in the middle layers and sialon and other hard phases such as alpha alumina along the coating. The size of nanocrystallites in the coating was calculated around 71 nm. The mass loss rate increased by increasing of duty cycle and current density while it decreased by increasing of frequency and coating time. The volcano-like morphology of the final nanocomposite coating showed the local flow of molten oxide toward the outer surface of the coating and agglomeration of the nanoparticles in the valleys. The effects of frequency and duty cycle on the mass loss rate were more than that of current density. By this ranking, "time" of coating is placed in fourth level. Also, higher concentrations of silicon nitride nanoparticles in the suspension could embed more nanoparticles into the coating. Wear began on the first layer by partial removal of the layer on different distributed local zones on the surface. These local zones consist of possible agglomerated nanoparticles. Inner layers were removed slightly during the wear test and showed uniform worn surfaces. Layer 3 exhibited better wear conditions with respect to the other layers at low concentrations of nanoparticles. An increase in nanoparticle concentration led to a shift of this optimum layer toward the substrate.

REFERENCES

[1] Abad, M.D., Cuceres, D., Pogozhev, Y.S., Shtansky, D.V. and Sunchez-Lopez, J.C. Bonding structure and mechanical properties of Ti-B-C coatings. *Plasma Processes and Polymers*, 2009, 6(SUPPL. 1), S107-S112.

[2] Akbari, A., Riviere, J.P., Templier, C. and Le Bourhis, E. Structural and mechanical properties of IBAD deposited nanocomposite Ti-Ni-N coatings. *Surface and Coatings Technology*, 2006, 200(22-23 SPEC. ISS.), 6298-6302.

[3] Aliofkhazraei, M., Ahangarani, S. and Sabour Rouhaghdam, A. Effect of the duty cycle of pulsed current on nanocomposite layers formed by pulsed electrodeposition. *Rare Metals*, 2010, 29(2), 209-213.

[4] Aliofkhazraei, M., Ahangarani, S.H. and Rouhaghdam, A.S. Effect of surface nanocrystallization and PPEC time on complex nanocrystalline hard layer fabricated by plasma electrolysis. *Transactions of Nonferrous Metals Society of China*, 2010, 20(3), 425-431.

[5] Aliofkhazraei, M., Morillo, C., Miresmaeili, R. and Sabour Rouhaghdam, A. Carburizing of low-melting-point metals by pulsed nanocrystalline plasma electrolytic carburizing. *Surface and Coatings Technology*, 2008, 202(22-23), 5493-5496.

[6] Aliofkhazraei, M., Rouhaghdam, A.S., Ghobadi, E. and Mohsenian, E. Electrodeposition and mechanical and corrosion resistance properties of tertiary Ni-W/Al_2O_3/CNT nanocomposite coatings. *Advanced Materials Research*, 2010 (89-91), 12-16.

[7] Aliofkhazraei, M. and Sabour Rouhaghdam, A. Fabrication of TiC/WC ultra hard nanocomposite layers by plasma electrolysis and study of its characteristics. *Surface and Coatings Technology*, 2010.

[8] Aliofkhazraei, M., Sabour Rouhaghdam, A. and Shahrabi, T. Abrasive wear behaviour of Si_3N_4/TiO_2 nanocomposite coatings fabricated by plasma electrolytic oxidation. *Surface and Coatings Technology*, 2010.

[9] Aliov, M.K. and Sabur, A.R. Formation of a novel hard binary SiO_2/quantum dot nanocomposite with predictable electrical conductivity. *Modern Physics Letters* B, 2010, 24(1), 89-96.

[10] Azami, M., Moztarzadeh, F. and Tahriri, M. Preparation, characterization and mechanical properties of controlled porous gelatin/hydroxyapatite nanocomposite through layer solvent casting combined with freeze-drying and lamination techniques. *Journal of Porous Materials,* 2009, 1-8.

[11] Basnyat, P., Luster, B., Kertzman, Z., Stadler, S., Kohli, P., Aouadi, S., Xu, J., Mishra, S.R., Eryilmaz, O.L. and Erdemir, A. Mechanical and tribological properties of CrAlN-Ag self-lubricating films. *Surface and Coatings Technology*, 2007, 202(4-7), 1011-1016.

[12] Beake, B.D., Vishnyakov, V.M., Valizadeh, R. and Colligon, J.S. Influence of mechanical properties on the nanoscratch behaviour of hard nanocomposite TiN/Si_3N_4 coatings on Si. *Journal of Physics D: Applied Physics,* 2006, 39(7), 1392-1397.

[13] Benkahoul, M., Robin, P., Gujrathi, S.C., Martinu, L. and Klemberg-Sapieha, J.E. Microstructure and mechanical properties of Cr-Si-N coatings prepared by pulsed reactive dual magnetron sputtering. *Surface and Coatings Technology*, 2008, 202(16), 3975-3980.

[14] Bobzin, K., Bagcivan, N., Immich, P., Bolz, S., Cremer, R. and Leyendecker, T. Mechanical properties and oxidation behaviour of (Al,Cr)N and (Al,Cr,Si)N coatings for cutting tools deposited by HPPMS. *Thin Solid Films*, 2008, 517(3), 1251-1256.

[15] Borcea, B., Munteanu, A., Munteanu, D., Olteanu, C., Guilaumont, A. and Klein, D. Mechanical properties of the nanocomposite TI-SI-N thin films deposited by magnetron sputtering using a HIPIMS/DC pulsed device. *Metalurgia International*, 2009, 14(SPEC. ISS. 3), 125-128.

[16] Bouchet, J., Rochat, G., Leterrier, Y., M¥nson, J.A.E. and Fayet, P. The role of the amino-organosilane/SiO_x interphase in the barrier and mechanical performance of nanocomposites. *Surface and Coatings Technology*, 2006, 200(14-15), 4305-4311.

[17] Bouzakis, K.D., Skordaris, G., Michailidis, N., Mirisidis, I., Erkens, G. and Cremer, R. Effect of film ion bombardment during the pvd process on the mechanical properties and cutting performance of TiAlN coated tools. *Surface and Coatings Technology*, 2007, 202(4-7), 826-830.

[18] Carvalho, S., Ribeiro, E., Rebouta, L., Tavares, C., Mendonc§a, J.P., Caetano Monteiro, A., Carvalho, N.J.M., De Hosson, J.T.M. and Cavaleiro, A. Microstructure, mechanical properties and cutting performance of superhard (Ti,Si,Al)N nanocomposite films grown by d.c. reactive magnetron sputtering. *Surface and Coatings Technology*, 2004, 177-178, 459-468.

[19] Chang, Y.Y., Yang, S.J. and Wang, D.Y. Structural and mechanical properties of Cr-C-O thin films synthesized by a cathodic-arc deposition process. *Surface and Coatings Technology*, 2007, 202(4-7), 941-945.

[20] Chen, L., Du, Y., Wang, A.J., Wang, S.Q. and Zhou, S.Z. Effect of Al content on microstructure and mechanical properties of Ti-Al-Si-N nanocomposite coatings. *International Journal of Refractory Metals and Hard Materials*, 2009, 27(4), 718-721.
[21] Choi, S.R., Park, I.W., Kim, S.H. and Kim, K.H. Effects of bias voltage and temperature on mechanical properties of Ti-Si-N coatings deposited by a hybrid system of arc ion plating and sputtering techniques. *Thin Solid Films*, 2004, 447-448, 371-376.
[22] Dearnley, P.A., Kern, E. and Dahm, K.L. Wear response of crystalline nanocomposite and glassy Al_2O_3-SiC coatings subjected to simulated piston ring/cylinder wall tests. Proceedings of the Institution of Mechanical Engineers, Part L: *Journal of Materials: Design and Applications*, 2005, 219(2), 121-137.
[23] Ding, X.Z., Zeng, X.T., Liu, Y.C., Yang, Q. and Zhao, L.R. Structure and mechanical properties of Ti-Si-N films deposited by combined DC/RF reactive unbalanced magnetron sputtering. *Journal of Vacuum Science and Technology A: Vacuum, Surfaces and Films*, 2004, 22(6), 2351-2355.
[24] Druffel, T., Geng, K. and Grulke, E. Mechanical comparison of a polymer nanocomposite to a ceramic thin-film anti-reflective filter. *Nanotechnology*, 2006, 17(14), 3584-3590.
[25] Fragneaud, B., Masenelli-Varlot, K., Gonzalez-Montiel, A., Terrones, M. and Cavaill, J.Y. Mechanical behavior of polystyrene grafted carbon nanotubes/polystyrene nanocomposites. *Composites Science and Technology*, 2008, 68(15-16), 3265-3271.
[26] Galvan, D., Pei, Y.T. and De Hosson, J.T.M. Influence of deposition parameters on the structure and mechanical properties of nanocomposite coatings. *Surface and Coatings Technology*, 2006, 201(3-4), 590-598.
[27] Guruvenket, S., Li, D., Klemberg-Sapieha, J.E., Martinu, L. and Szpunar, J. Mechanical and tribological properties of duplex treated TiN, nc-TiN/a-SiN_x and nc-TiCN/a-SiCN coatings deposited on 410 low alloy stainless steel. *Surface and Coatings Technology*, 2009, 203(19), 2905-2911.
[28] Han, D.S., Song, P.K., Cho, K.M., Park, Y.H. and Kim, K.H. Synthesis and mechanical properties of Ti-Si-C films by a plasma-enhanced chemical vapor deposition. *Surface and Coatings Technology*, 2004, 188-189(1-3 SPEC.ISS.), 446-451.
[29] Hong, S.G., Shin, D.W. and Kim, K.H. Syntheses and mechanical properties of quaternary Cr-Mo-Si-N coatings by a hybrid coating system. *Materials Science and Engineering* A, 2008, 487(1-2), 586-590.
[30] Hou, F., Wang, W. and Guo, H. Effect of the dispersibility of ZrO2 nanoparticles in Ni-ZrO_2 electroplated nanocomposite coatings on the mechanical properties of nanocomposite coatings. *Applied Surface Science,* 2006, 252(10), 3812-3817.
[31] Ivashchenko, V.I., Porada, O.K., Ivashchenko, L.A., Timofeeva, I.I., Dub, S.M. and Skrinskii, P.L. Mechanical and tribological properties of TiN and SiCN nanocomposite coatings deposited using methyltrichlorosilane. *Powder Metallurgy and Metal Ceramics*, 2008, 47(1-2), 95-101.
[32] Jacquelot, E., Galy, J., Gerard, J.F., Roche, A., Chevet, E., Fouissac, E. and Verch¦¨re, D. Morphology and thermo-mechanical properties of new hybrid coatings based on polyester/melamine resin and pyrogenic silica. *Progress in Organic Coatings*, 2009, 66(1), 86-92.

[33] Jeon, Y.S., Byun, J.Y. and Oh, T.S. Electrodeposition and mechanical properties of Ni-carbon nanotube nanocomposite coatings. *Journal of Physics and Chemistry of Solids*, 2008, 69(5-6), 1391-1394.

[34] Kertzman, Z., Marchal, J., Suarez, M., Staia, M.H., Filip, P., Kohli, P. and Aouadi, S.M. Mechanical, tribological, and biocompatibility properties of ZrN-Ag nanocomposite films. *Journal of Biomedical Materials Research* - Part A, 2008, 84(4), 1061-1067.

[35] Khazrayie, M.A. and Aghdam, A.R.S. Si_3N_4/Ni nanocomposite formed by electroplating: Effect of average size of nanoparticulates. *Transactions of Nonferrous Metals Society of China*, 2010, 20(6), 1017-1023.

[36] Khazrayie, M.A. and Aghdam, A.S.R. Characterization of Ni-W/MWCNT nanocomposite layers formed by pulsed electrochemical deposition. *Protection of Metals*, 2010(6).

[37] Kim, C., Kang, M.C., Kim, J.S., Kim, K.H., Shin, B.S. and Je, T.J. Mechanical properties and cutting performance of nanocomposite Cr-Si-N coated tool for green machining. *Current Applied Physics*, 2009, 9(1 SUPPL.), S145-S148.

[38] Kim, S.H., Jang, J.W., Kang, S.S. and Kim, K.H. Synthesis and mechanical evaluation of nanocomposite coating layer of nc-TiN/a-Si_3N_4 on SKD 11 steel by sputtering. *Journal of Materials Processing Technology*, 2002, 130-131, 283-288.

[39] Kuo, Y.C., Lee, J.W., Wang, C.J. and Chang, Y.J. The effect of Cu content on the microstructures, mechanical and antibacterial properties of Cr-Cu-N nanocomposite coatings deposited by pulsed DC reactive magnetron sputtering. *Surface and Coatings Technology*, 2007, 202(4-7), 854-860.

[40] Lee, J.W. and Chang, Y.C. A study on the microstructures and mechanical properties of pulsed DC reactive magnetron sputtered Cr-Si-N nanocomposite coatings. *Surface and Coatings Technology*, 2007, 202(4-7), 831-836.

[41] Lee, J.W., Kuo, Y.C. and Chang, Y.C. Microstructure and mechanical properties of pulsed DC magnetron sputtered nanocomposite Cr-Cu-N thin films. *Surface and Coatings Technology*, 2006, 201(7 SPEC. ISS.), 4078-4082.

[42] Li, J., Sun, Y., Sun, X. and Qiao, J. Mechanical and corrosion-resistance performance of electrodeposited titania - Nickel nanocomposite coatings. *Surface and Coatings Technology*, 2005, 192(2-3), 331-335.

[43] Li, J.Y., Dai, H., Zhong, X.H., Zhang, Y.F., Ma, X.F., Meng, J. and Cao, X.Q. Effect of the addition of YAG ($Y_3Al_5O_{12}$) nanopowder on the mechanical properties of lanthanum zirconate. *Materials Science and Engineering* A, 2007, 460-461, 504-508.

[44] Louro, C., Lamni, R. and Levy, F. W-B-N sputter-deposited thin films for mechanical application. *Surface and Coatings Technology*, 2005, 200(1-4 SPEC. ISS.), 753-759.

[45] Lu, C., Donch, I., Nolte, M. and Fery, A. Au nanoparticle-based multilayer ultrathin films with covalently linked nanostractures: Spraying layer-by-layer assembly and mechanical property characterization. *Chemistry of Materials*, 2006, 18(26), 6204-6210.

[46] Lv, H., Zhao, W., An, Q., Nie, P., Wang, J. and Chu, P.K. Nanomechanical properties and microstructure of ZrO_2/Al_2O_3 plasma sprayed coatings. *Materials Science and Engineering* A, 2009, 518(1-2), 185-189.

[47] Madhavan, K., Gnanasekaran, D. and Reddy, B.S.R. Synthesis and characterization of poly(dimethylsiloxaneurethane) nanocomposites: Effect of (in)completely condensed

silsesquioxanes on thermal, morphological, and mechanical properties. *Journal of Applied Polymer Science,* 2009, 114(6), 3659-3667.

[48] Mammeri, F., Le Bourhis, E., Rozes, L. and Sanchez, C. Elaboration and mechanical characterization of nanocomposites thin films. Part I: Determination of the mechanical properties of thin films prepared by in situ polymerisation of tetraethoxysilane in poly(methyl methacrylate). *Journal of the European Ceramic Society*, 2006, 26(3), 259-266.

[49] Mei, F., Shao, N., Hu, X., Li, G. and Gu, M. Microstructure and mechanical properties of reactively sputtered Ti-Si-N nanocomposite films. *Materials Letters*, 2005, 59(19-20), 2442-2445.

[50] Mercs, D., Bonasso, N., Naamane, S., Bordes, J.M. and Coddet, C. Mechanical and tribological properties of Cr-N and Cr-SI-N coatings reactively sputter deposited. *Surface and Coatings Technology*, 2005, 200(1-4 SPEC. ISS.), 403-407.

[51] Mirzamohammadi, S., Aliov, M.K., Sabur, A.R. and Hassanzadeh-Tabrizi, A. Study Of Wear Resistance And Nanostructure For Tertiary Al_2O_3/Y_2O_3/CNT Pulsed Electrodeposited Ni-Based Nanocomposite. *Materials Science*, 2010(1).

[52] Mirzamohammadi, S., Kiarasi, R., Aliov, M.K., Sabur, A.R. and Hassanzadeh-Tabrizi, A. Study of corrosion resistance and nanostructure for tertiary Al_2O_3/Y_2O_3/CNT pulsed electrodeposited Ni based nanocomposite. *Transactions of the Institute of Metal Finishing*, 2010, 88(2), 93-99.

[53] Mitterer, C., Lechthaler, M., Gassner, G., Fontalvo, G.A., Tooth, L., Picz, B., Raible, M., Maier, K. and Bergmann, E. Self-lubricating chromium carbide/amorphous hydrogenated carbon nanocomposite coatings: A new alternative to tungsten carbide/amorphous hydrogenated carbon. Proceedings of the Institution of Mechanical Engineers, Part J: *Journal of Engineering Tribology*, 2009, 223(5), 751-757.

[54] Nemeth, S. and Liu, Y.C. Mechanical properties of hybrid sol-gel derived films as a function of composition and thermal treatment. *Thin Solid Films*, 2009, 517(17), 4888-4891.

[55] Oliker, V.E., Sirovatka, V.L., Timofeeva, I.I., Gridasova, T.Y. and Hrechyshkin, Y.F. Formation of detonation coatings based on titanium aluminide alloys and aluminium titanate ceramic sprayed from mechanically alloyed powders Ti-Al. *Surface and Coatings Technology*, 2006, 200(11), 3573-3581.

[56] Park, I.W., Choi, S.R., Suh, J.H., Park, C.G. and Kim, K.H. Deposition and mechanical evaluation of superhard Ti-Al-Si-N nanocomposite films by a hybrid coating system. *Thin Solid Films*, 2004, 447-448, 443-448.

[57] Paternoster, C., Fabrizi, A., Cecchini, R., Spigarelli, S., Kiryukhantsev-Korneev, P.V. and Sheveyko, A. Thermal evolution and mechanical properties of hard Ti-Cr-B-N and Ti-Al-Si-B-N coatings. *Surface and Coatings Technology,* 2008, 203(5-7), 736-740.

[58] Polychronopoulou, K., Rebholz, C., Baker, M.A., Theodorou, L., Demas, N.G., Hinder, S.J., Polycarpou, A.A., Doumanidis, C.C. and Bubel, K. Nanostructure, mechanical and tribological properties of reactive magnetron sputtered TiCx coatings. *Diamond and Related Materials,* 2008, 17(12), 2054-2061.

[59] Pramanik, M., Srivastava, S.K., Samantaray, B.K. and Bhowmick, A.K. EVA/Clay Nanocomposite by Solution Blending: Effect of Aluminosilicate Layers on Mechanical and Thermal Properties. *Macromolecular Research*, 2003, 11(4), 260-266.

[60] Prasad, S.V., Scharf, T.W., Kotula, P.G., Michael, J.R. and Christenson, T.R. Application of diamond-like nanocomposite tribological coatings on LIGA microsystem parts. *Journal of Microelectromechanical Systems,* 2009, 18(3), 695-704.
[61] Reddy, C.S. and Das, C.K. Propylene-ethylene copolymer filled nanocomposites: Influence of Zn-ion coating upon nano-SiO_2 on structural, thermal, and dynamic mechanical properties. *Polymer - Plastics Technology and Engineering,* 2006, 45(7), 815-820.
[62] Sahoo, N.G., Jung, Y.C., Yoo, H.J. and Cho, J.W. Influence of carbon nanotubes and polypyrrole on the thermal, mechanical and electroactive shape-memory properties of polyurethane nanocomposites. *Composites Science and Technology,* 2007, 67(9), 1920-1929.
[63] Santana, A.E., Karimi, A., Derflinger, V.H. and Schuꞔ^tze, A. Thermal treatment effects on microstructure and mechanical properties of TiAlN thin films. *Tribology Letters*, 2004, 17(4), 689-696.
[64] Schmidt, D.J., Cebeci, F.İ., Kalcioglu, Z.I., Wyman, S.G., Ortiz, C., Van Vliet, K.J. and Hammond, P.T. Electrochemically controlled swelling and mechanical properties of a polymer nanocomposite. *ACS Nano,* 2009, 3(8), 2207-2216.
[65] Schwaller, P., Haug, F.J., Michler, J. and Patscheider, J. Nanocomposite hard coatings: Deposition issues and validation of their mechanical properties. *Advanced Engineering Materials*, 2005, 7(5), 318-322.
[66] Sharma, S.P. and Lakkad, S.C. Anchoring effect on the mechanical properties of CNTs grown carbon fiber/polymer matrix multi-scale composites. *Current Nanoscience*, 2009, 5(3), 306-311.
[67] Shi, L., Sun, C., Gao, P., Zhou, F. and Liu, W. Mechanical properties and wear and corrosion resistance of electrodeposited Ni-Co/SiC nanocomposite coating. *Applied Surface Science*, 2006, 252(10), 3591-3599.
[68] Shokuhfar, T., Makradi, A., Titus, E., Cabral, G., Ahzi, S., Sousa, A.C.M., Belouettar, S. and Gracio, J. Prediction of the mechanical properties of hydroxyapatite/polymethyl methacrylate/carbon nanotubes nanocomposite. *Journal of Nanoscience and Nanotechnology*, 2008, 8(8), 4279-4284.
[69] Sun, J., Francis, L.F. and Gerberich, W.W. Mechanical properties of polymer-ceramic nanocomposite coatings by depth-sensing indentation. *Polymer Engineering and Science*, 2005, 45(2), 207-216.
[70] Sun, Y., Sun, J., Liu, M. and Chen, Q. Mechanical strength of carbon nanotube-nickel nanocomposites. *Nanotechnology,* 2007, 18(50).
[71] Suszko, T., Gulbiski, W. and Jagielski, J. Mo_2N/Cu thin films - the structure, mechanical and tribological properties. *Surface and Coatings Technology*, 2006, 200(22-23 SPEC. ISS.), 6288-6292.
[72] Tseng, C.C., Hsieh, J.H., Wu, W., Chang, S.Y. and Chang, C.L. Surface and mechanical characterization of TaN-Ag nanocomposite thin films. *Thin Solid Films,* 2008, 516(16), 5424-5429.
[73] Wang, W., Hou, F.Y. and Guo, H.T. Relationship between dispersibility of ZrO_2 nanoparticles in Ni-ZrO_2 electroplated nanocomposite coatings and mechanical properties of nanocomposite coatings. *Transactions of Nonferrous Metals Society of China* (English Edition), 2004, 14(SUPPL. 2), 186-189.

[74] Wu, F.B., Tien, S.K., Lee, J.W. and Duh, J.G. Comparison in microstructure and mechanical properties of nanocomposite CrWN and nanolayered CrN/WN coatings. *Surface and Coatings Technology*, 2006, 200(10 SPEC. ISS.), 3194-3198.

[75] Xu, Y., Li, L.H., Cai, X., Chen, Q.L. and Chu, P.K. Microstructure and mechanical properties of Ti-Si-N nanocomposite coating prepared by DC magnetron sputtering. Shanghai Jiaotong Daxue Xuebao/*Journal of Shanghai Jiaotong University*, 2007, 41(3), 452-456.

[76] Yao, Y., Yao, S., Zhang, L. and Wang, H. Electrodeposition and mechanical and corrosion resistance properties of Ni-W/SiC nanocomposite coatings. *Materials Letters*, 2007, 61(1), 67-70.

[77] Yao, Y.W., Yao, S.W. and Zhang, L. Preparation, mechanical properties and wear resistance of Ni-W/SiC nanocomposite coatings. *Materials Science and Technology,* 2008, 24(2), 237-240.

[78] Zehnder, T., Schwaller, P., Munnik, F., Mikhailov, S. and Patscheider, J. Nanostructural and mechanical properties of nanocomposite nc-TiC/a-C:H films deposited by reactive unbalanced magnetron sputtering. *Journal of Applied Physics*, 2004, 95(8), 4327-4334.

[79] Zhang, C.H., Lu, X.C., Wang, H., Luo, J.B., Shen, Y.G. and Li, K.Y. Microstructure, mechanical properties, and oxidation resistance of nanocomposite Ti-Si-N coatings. *Applied Surface Science*, 2006, 252(18), 6141-6153.

[80] Zhang, P., Cai, Z. and Xiong, W. Influence of Si content and growth condition on the microstructure and mechanical properties of Ti-Si-N nanocomposite films. *Surface and Coatings Technology*, 2007, 201(15), 6819-6823.

[81] Zhang, X.D., Meng, W.J., Wang, W., Rehn, L.E., Baldo, P.M. and Evans, R.D. Temperature dependence of structure and mechanical properties of Ti-Si-N coatings. *Surface and Coatings Technology*, 2004, 177-178, 325-333.

[82] Zhang, Z.G., Rapaud, O., Allain, N., Baraket, M., Dong, C. and Coddet, C. Structure and mechanical properties of nanoscale multilayered CrN/ZrSiN coatings. *Journal of Vacuum Science and Technology A: Vacuum, Surfaces and Films,* 2009, 27(4), 672-680.

[83] Ziebert, C., Albers, U., Stuber, M. and Ulrich, S. Constitution and mechanical properties of nanocrystalline reactive magnetron sputtered V-Al-C-N hard coatings as a function of the carbon content. *Plasma Processes and Polymers*, 2009, 6(SUPPL. 1), S560-S565.

[84] Zou, C.W., Wang, H.J., Li, M., Yu, Y.F., Liu, C.S., Guo, L.P. and Fu, D.J. Microstructure and mechanical properties of Cr-Si-N nanocomposite coatings deposited by combined cathodic arc middle frequency magnetron sputtering. *Journal of Alloys and Compounds*, 2009, 485(1-2), 236-240.

Chapter 7

CONCLUSIONS

Based on the discussions in different chapters of the book, these conclusions and suggestions were extracted:

1) In surface engineering, considered requirements for surface are: increasing the strength against friction, abrasion, and corrosion, or boosting the thermal, optical, magnetic, and electrical properties. For instance, fuel yield and specific output power of thermal engines, such as gas turbines or adiabatic diesels, is limited by hot corrosion and the properties of thermal barriers of special pieces of surfaces. Similarly, in a wide range of industry (such as nuclear power, gas and oil exploitation, mining, and manufacturing), the surface generates an important problem in technological advancement. Using total material for improving surface properties is not economically advised. Thus, for high rate of yield it is recommended to use a sub-layer with efficient properties, and materials which are cheaper and easier to reshape. Ideally, sub-layer must be optimized for maximizing the coat benefits and, consequently, creating the most efficient coating system. Hence, the final product eliminates the need for rare materials and can be a cheaper and of a better yield solution, compared with early solutions.
2) Electroplating and electrodeposition are performed through electrical deposition process, which is indeed a metal coating deposit created by electrical current application. At electrical deposition, negative charge is applied on covered block surface, using an external voltage. Then coatings are developed by submerging the pieces into a solution - rich by metallic ions - and applying the electrical current through that. Metallic ions are provided by either metallic salts dissolution into plating solution (electrolyte) or positive pole (anodes) electrode dissolution during plating. In this condition, due to potential difference, positive ions (cations) move toward cathode and reduce on its surface as metallic atoms. At electrical deposition there must be an electrical circuit, in the presence of a battery or supply resource – far from charged ions and electrolyte in solution. Thus, for performing electrical deposition process the circuits' electrical conductivity and metallic ions motivation is necessary. Applying the electrical conductivity of circuit, sub-layer's surface must include no surface pollutions or oxides. In this case metallic ions mobility increases through heating and lack of impurities in electrolyte.

3) Advantages of electroplating are: (a) As operation temperature does not exceed over 100 °C, work-piece is not subject to any metallurgic deformation or unwanted changes. (b) Adjusting the plating condition one can change the solidity, internal stress and metallurgic characteristic of the coating. (c) The obtained coating through this method has an efficient compaction and adhesion with its sub-layer. In this method the bound nature is molecular and might reach up to 1000 MPa. (d) The coating thickness is adjustable through a change in applied current density and the length of plating. (e) In this case, there are no technical limitations in coating thickness; and for metals such as nickel, we can obtain the thickness of 13mm by electroforming and reclamation. (f) Creating this kind of coating, in some extent, is not affected by line of sight factor. Although the throwing power (i.e. the ability of coating at rounded corners) may be limited, but there is a relative freedom in anodes arranging (e.g. in slim tubes interior coating). (g) We can easily cover the surfaces which has no need for coating. (h) The process is efficient for automation.
4) Disadvantages of electroplating are: (a) As current density distribution is not equal in the piece, the coating intends to be thickened at corners and edges and be thinned at dents and center of flat broad surfaces. (b) Deposition rate rarely exceeds over 75μm/h; however through electrolyte circulation it can be accelerated. (c) The size of plating bath is limited by work-piece dimensions.
5) Composite plating, which is also addressed as electrochemical composite plating, is among the methods for creating composite materials with metallic matrix. Composite plating is carried out via simultaneous electrodeposition process, which is a process of combining the insoluble dispersive particles in electrolyte during an electrical deposition of a metal. Through this process we can develop thin layers of metal matrix composites bearing particles of pure metal, ceramics and organic materials. The particles sizes are from 2 nm to 100μm, which are generally used in the matrix of copper, nickel, cobalt, and different alloys, resulted from electrodeposition. Suspended particles concentration in plating solution varies between 2 to 200 g/lit; and one can typically produce composites with volumetric ratio of 1 to 50 from them.
6) One of early models, seen in many of articles, is offered by Guglielmi. This model suggests that particles absorbed charge, ζ, is the main factor in determining feasibility of simultaneous deposition. The model is based on two successive adsorption stages. The first stage is the transition of a layer from weak absorbed particles (with relatively high coverage range compared with Helmoltz double layer), which occurs due to intense agitation. At the second stage, first charged particles are transferred at high potential gradient through the electrophoresis adsorption properties on cathode surface, and then particles are absorbed on cathode surface due to coulomb's force between particle and absorbed anions – which causes particles strongly to be absorbed on electrode.
7) There is a more advanced model, with five steps. This model is based on this assumption that the only time for a particle to be accommodated in the coating is when a determined fragment of ionic clouds ions are reduced. Since most materials be charged in contact with an aquatic (polar) environment, first stage does not seem necessary. In this state, anions such as OH^- surround particle surface and develop a dual electric layer. At the second stage, particle transfer to cathode surface might be done through mechanisms such as displacement, penetration, or electrophoresis,

which are basically rely on particle sizes and the applied forces on particles. The applied forces on the particle can be divided into hydrodynamic (resulted from liquids movement through electrolyte agitation) forces, and forces such as gravity, buoyancy, and interactive forces between particle and electrode. For particles of nanometer size, the mechanism is the connective diffusion; while in particles of larger sizes – in micrometer ranges – mobility and gravity forces are more important. For electrical migration, the electrical charge on particles' surface or zeta (ζ) potential seems significant. Adding positive ions such as Tl^+, Cs^+, and NH_4^+, (called cathionic additives or surface active agents) we can raise particle's positive ion.

8) In colloidal systems, where the dispersive phase or colloid has at least one dimension in nanometer to micrometer ranges, colloidal stability or dispersion is considered as an important issue. Generally, the random collisions between dispersive particles take place in a liquid environment. Since composite plating electrolytes are examples of colloidal systems, there is even a chance of insolvable particles adhesion. When these particles stick to electrolytes, an increase in the number of adhered particles in composite's coatings will happen. Therefore, the unique attribution of composite coatings - particularly in the presence of nanoparticles which stick easily to each other due to a high surface energy – will be removed and is considered as an unfavorable factor. Stability of dispersion or particles inclining to agglomeration is defined through particles interaction during the collision. This interaction includes attraction of van der Waals forces towards electrostatic repellent forces, is resulted from overlapping of particles' dual layer.
9) Monovalent cationic additives such as Tl^+, Cs^+, and NH_4^+ get easily absorbed by particles and enhance electrostatic repellent forces among particles through modification of particles surface electrical charge. Hence particles agglomeration will take places in the electrolyte with low abundance. On one hand these ions attraction, as it previously mentioned, conducts in accelerating of electrical migration in suspended particles of electrolyte. On the other hand simultaneous deposition of particles which are weakly attached to cathode increases on cathode surface – due to their reduction. In this state if cationic additives are the main absorbed ions on particles surface, they might create bounds between particles and matrix phase and even act as spots for electro-crystallization initiation. On the other cases, cationic additives can make a favorable condition for absorbing matrix phase ions on particles surface; since matrix phase ions would be main absorbed category. At this state, there will be an efficient bound between particles and matrix phase and matrix phases' formation will increase. In both states mentioned above cationic additives reduction might strongly prevent growth of matrix phase grains and leads to development of equi-axial grains.
10) Once surfactants exposed to emulsion and suspension solutions and mixtures, they exhibit a great tendency to initiate adsorption interfaces. Thus, it may cause to a decrease in surface tension, wetting, and dispersion. Surfactants are an organic molecule with a part which solvable in water (hydrophilic) and a part which is solvable in lipid (hydrophobic). Regarding surfactants structure they can be classified into four groups: anionic, cationic, non-ionic, and amphoteric. In anionic surfactants the hydrophilic part has negative electrical charge, while the cationic ones have a positive charge. Non-ionic surfactants have no charge and amphoteric ones have both

positive and negative charges on each part of molecule. In composite plating mostly cationic and non-ionic surfactants are used. Surfactants cause particles dispersion through enhancing their surface electrical potential (electrical charge). Applying surfactants with aromatic cycles (enriched with electron) and with a flat molecular structure, for their easy adsorption and making an organic barrier around particles, we can significantly (or even completely) avoid particles agglomeration. At surfactants with very low concentration the solution seems to be transparence with buoyant particles. In solutions with a little higher concentration, although surfactants are of a little concentration, the solution seems to be blurry or cloudy. As we increase the surfactants ratio, the solution gradually turns into beige colored one with colloidal stability requirement. In surfactants' higher concentrations the solution will be completely transparent.

11) It seems that the internal stress of electrochemical composite coatings depends on particles type, amount, and size. For instance, adding organic materials, such as saccharine decreases the nickel coatings' internal stress from Watts' solution, and converts it from tensional to compressive. On the other hand it was said that, once the particles amount is high there will be a strong tensional stress in the coating. Some researchers believed that micro and macro stresses are caused by a discrepancy between particles mechanical properties and coatings' and recorded that in a nickel coating - in presence of Si_3N_4 particles – for particles with 5.2 μm, as particles number raises, coating's tensional stress will boost too; while for 0.8 μm and 3.6μm particles an increase in particles number leads to a decrease in coatings tensional stress. In this aspect some researches implies that alumina and titanium particles presence, with average of 13 nm and 21 nm diameter, causes a decrease in remained stress in nickel coating.
12) A good choice of ceramic materials for producing ceramic composites is of a great importance. Non-oxide ceramics, such as SiC and Si_3N_4, have some significant mechanical properties at high temperatures; however they will oxidize at temperatures over 1500 °C in air proximity. Although oxide ceramics enjoy a better strength against oxidation and can be better sintered than non-oxide ones, but they have a lower creep strength which limits their applications. For example, alumina is one mostly used engineering ceramics with many usages in different industries. It is also applied in high temperatures and shows slight creep properties, being subjected to stresses. For their creep improvement a material with high creep strength from secondary phase will be used- or as one can say an alumina composite will be made. However different materials are used in alumina bearing composites, but among them YAG (Yttrium Aluminum Garnet, $Y_3Al_5O_{12}$) is one of the best choices for enhancing alumina creep strength – as a mono crystal of YAG has ten times greater creep strength than alumina. YAG's particular structure stops movements and displacements.
13) Nanotechnology is a concept which is addressed to all advance technologies in the field of work with nano scale. Usually when it is said nano scale, it means 1 to 100 nm. However, it must be added there are no distinct boundaries for this definition, as in some references 1 to 250 nanometers is considered in nano range. But the important thing is that there are some characteristics which are shown in nano scale materials, as it is expected that, materials with nano structures have better

mechanical, physical, and biological properties than that of materials with micro structures. In addition, in nano ranges powders can be better sintered. The first traces of nano-technology - however not known by this name - came back to 1959. In this year Richard Fineman during a lecture - called “there are much spaces in lower levels” - introduced the notion of nano-technology. Through this theory he announced that molecules and atoms can be directly manipulated in the near future. Nanotechnology gradually entered into sciences such as physics, chemistry, medicine, and, particularly, materials science.

14) Generally one can categorized nano composites into three groups. This classification is based on distributed phases type and apparent shape in matrix. There are three types of reinforcing materials, including layered, fibrous, and particulate. As these particles size decreases, their surface to volume ratio increases which can introduce new features to composites. Platelet uniform distribution is an important subject which must be considered. For this group we can name clay/polymer composites. In a typical state clay layers intervals are minimum. Once polymeric matrix diffuses into clay layers, layers intervals will increase. A polymeric matrix, also, can completely separate layers from each other. Each of these states can create different characteristics in composites. Applying clay/polymer nano-composites improves properties such as strength, hardness, ductility, and thermal stabilization. Nowadays, these composites are used in manufacturing many vehicle panels. The fibers can be different materials. One of materials which are highly interested during these days is carbon nano-tubes. For instance, producing a nano-composite with alumina matrix, researchers provide an alumina sol, using an aluminum Alco-oxide, and then alcohol and carbon nano-tube mixture is added to sol, which turns the sol into a gel. Finally, after calcination process, a composite of alumina-CNT will be obtained. Similar to previous ones, uniform distribution of secondary phase have a great effect on final nano-composites properties. In addition, the interface form between matrix and fibers is of a great importance; as more chemical boundaries (versus to a weak physical boundary) have higher impact on strength and ductility enhancement. In nanoparticle reinforced composites, secondary phases are distributed in matrix. May one can say these form of composites are more common than the others. They classified into three categories: (a) Intra-type nanostructure composites (b) Inter-type nanostructure composites (c) Nano/nano nanostructure composites

15) Sol-gel method is another chemical method for producing nano-materials. Sol-gel is a process for producing materials by developing sol, converting sol to gel, and finally solution exit. The solution can be each of Alco-oxide or salt combinations. Sol-gel is a chemical, or semi-chemical, method which is able to produce the desired product in a semi-stable (or amorphous) state, through creating a network by the relevant reactions. The point which gives priority for this method is the uniformity and homogeneity of the product - as one can say materials are combined with each other in molecular level. Sol’s stabilization or coalescence is important in sol-gel process, while the obtained gel network and structure is related to agglomerated particles size and shape. Accurate controlling of intra-particle forces, we can achieve a colloidal suspension in a stable, weak follicular, and strong follicular state. Colloidal suspension stabilized state is where their potential energy increases through particles approaching to each other, and consequently detract each other in the suspension. In

weak follicular state energy curve involves a minimum point. In this state agglomerated particles are in a balanced distance from each other, which in fact is the curves minimum distance.

16) The most important factors affecting particle distribution in the solution are surface charge of suspending particles and ionic solidity of the solution in which the particles are suspending. The quantity of the ratio of these two parameters determines how the particles interact with each other. When the ratio of surface charge of the particles and ionic solidity is high, the result of the interaction between the particles is repulsion, and consequently a low-viscosity solution. Conversely, when quantity of the ratio is low, the particles will be in agglomerated form inside the solution. The effect of ionic power of the solution on agglomeration of silicon carbide particles (250 nm) illustrated that agglomeration rate of silicon carbide particles in distilled water containing 0.1 g/l silicon carbide and without using additives, is much lower that in nickel electroplating bath (sulfamate bath) containing the same amount of silicon carbide under similar conditions. This observation is pertinent to the higher ionic power of the electroplating solution than that of water. When there are many equally-charged particles, they will be distributed in the solution. However, surface charge control through checking pH might not be sufficient to create a suspension of minimum agglomeration and maximum stability. Typically, the best practical way to create a stable suspension is using chemical stabilizers called scatters, which are usually polyelectrolyte surfactants that adsorb on particle surface, change its surface charge, and consequently alter the electrostatic forces.

17) Different nano-sized particles (in spherical form) in the range of 4 to 800 nm have been successfully deposited in metallic matrixes. Aluminum oxide (Al_2O_3), chromium (Cr), diamond (C), silicon carbide (SiC), gold (Au), Silicon oxide (SiO_2), Zirconium oxide (ZrO_2), Titanium oxide (TiO_2), polystyrene (PS), and silicon nitride (Si_3N_4) are the well-known examples. Among different metals, copper and nickel have been widely used, and alumina particles have been studied more than the other nano-metric particles. The entrance of nano-metric particles into the coating depends on many process parameters such as particle specifications (concentration, surface charge, type, form, and size), electrolyte composition (electrolyte concentration, additives, temperature, pH, and type of the surfactant), current density (direct current and pulse current), frequency, time percent of illumination, type of bath turbulence (layered, mixed, or turbulent), and electrode geometry (rotating disc, rotating cylinder, parallel sheets of the electrode, and many other conditions).

18) Particle size affects particle presence rate in the coating. Previous research has shown that more particles can enter a volume unit of the coating by reducing particle size. For instance, participation rates of alumina particles of two different sizes (50 nm and 300 nm) have been compared in nickel coating. In similar electroplating conditions, it has been proven that volume percentage ratio of 300 nm alumina particles in nickel coating is much higher than that of 50 nm alumina particles. Nonetheless, the number of 50 nm alumina particles in the coating is much larger than that of 300 nm alumina particles.

19) The main objectives of producing nano-composite coatings are to increase their hardness and resistance against wear and corrosion. Several terms must be in mind for studying nanocomposites: (1) nano-crystalline structures have higher micro-

hardness compared with microcrystal structures; (2) the coatings obtained from a pulse current are of higher hardness compared with the ones obtained from a direct current. Different reinforcing particles such as oxides, carbides, nitrides, sulfides, polymers, graphite and nano-fibers and carbon nano-tubes, metallic particles and diamond - depending on their particular characteristics - are added to metallic matrix. Regarding their expected properties and applications, these particles may have dimensions about several nano-meters (such as Al_2O_3 with average size of less than 14 nm in nickel matrix) to several micrometers (such as diamonds with average size of 150 microns particles in same matrix). These reinforcing particles may have one or more of mentioned characteristics. However, achieving desirable properties through one particle type may result in reducing another capability of the coating. For instance, adding Al_2O_3 and TiO_2 particles for enhancing nickel coatings' hardness and abrasion resistance leads to decrease of wear strength in a chlorine content environment, since nickel particle interface is a suitable path for diffusion of invader ions, such as chlorine ions, into the coating.

20) About simultaneous deposition of particle-metal it can be said that: at high current densities Ni solved ions quickly move from anode toward cathode. However, SiC larger particles which move through convectional mechanism, cannot reach to cathode surface at this speed. As a result the higher amount of Ni will be reduced against particles on the surface and content of particles within the surface will drop. On the other hand, at low current densities, Ni ions are slowly moving into the electrolyte thus there is no enough time for Ni cations to be absorbed. This leads to lack of particles deposition on cathode surface. SiC particles agglomeration at low densities is also due to this low speed of Ni cations movements and lack of full coverage of SiC particles surface with them. On the other hand, an unusual increase of current density conducts in increase of coatings internal stresses and, as a consequence, SiC particles rejection. With respect to all discussed points it seems that the required current density for obtaining maximum particle volumetric percentage into the coating has an optimum value. It may be suggested this optimum current density is influenced by some other factors such as bath agitation rate, size and amount of particles of the bath. However, there is no clear record for its amount and it just can be told the studies are generally performed using current of 1 to 10 A/dm^2.

21) Electrodeposition of composite coatings includes some type of electroplating where some particles in micron or less than micron dimensions are suspended in electrolyte and they enter into coating during the coating process. Electrodeposition in manufacture of composite coatings includes electroplating and electrolysis techniques. In electrolysis method, metallic coating deposit are generated on the surface of piecework by chemical reactions, and the dominant mechanism is to particles strong attraction which covered by reduction of appropriate quantity of metallic complex ions at the level of existing particles over cathode surface. However in electroplating, metallic coating is deposited by using external current and reduction of metallic ions appropriate quantity over cathode surface and the main mechanism is strong absorption of reduced particles which are absorbed weakly by cathode. Of electrolysis method advantages, one could imply further hardness and uniformity, and of electrodeposition technique advantages to further ductility and

electrolyte lower cost and more speed in coating process etc. One could use direct and pulsed current in production of nanocomposite coatings by electroplating technique. In pulsed electroplating, one may use different forms like triangular, saw tooth, and squared pulses etc. In pulsed electroplating technique, greater amount of particles with higher uniformity enter into the matrix than in direct electroplating, and on the other hand, due to presence of more parameters in production process, it is possible to control coating properties in this technique. Production parameters like current density and operational cycle, frequency, solution revolving velocity and additive have highly affect on the quantity of strengthening particles as well as their propagation in the matrix.

22) The main objective in fabrication of some nanocomposite coatings is to increase in erosive resistance, hardness and resistance against corrosion. Many results in studies indicate that the presence of ceramic particles in coating may highly increase the resistance against erosion and corrosion. The reason for increasing resistance against erosion is attributed to factors like increase in hardness, and decrease in adhesive erosion between overlapping surfaces and adjustment of microstructure. The important point which should be considered in study of erosive properties of this category of materials is in addition to quantity and propagation of particles, they also play a significant role in resistance against erosion. Fine and homogenous propagation of particles causes improvement in erosive properties.
23) The hardness of commercial pure nickel deposit is about 215 to 265 Vickers while composite coating hardness varies from 275 to 850 Vickers. In fact, the presence of the particles in this coating may increase hardness at this coating. It has been characterized that composite coating hardness and other mechanical properties not only depend on mechanical properties of matrix and particles, but also on quantity and size of particles. Quantity and size of particles in both mechanisms may determine reinforcement in metallic matrix composites. It was found that coating hardness may increase by increase in the quantity of coating particles.

INDEX

D

E

F

G

H

T

U

V

W

X

Y

Z

INFORMACIÓN IMPORTANTE: Si has recibido previamente un correo electrónico deberás seguir los pasos que en él se detallan.

Estimado/a cliente/a,

Para acceder a la versión electrónica de este libro, por favor, accede a **http://onepass.aranzadi.es** Tras acceder a la página citada, introduce tu dirección de correo electrónico (*) y el código que encontrarás en el interior de la cubierta del libro.

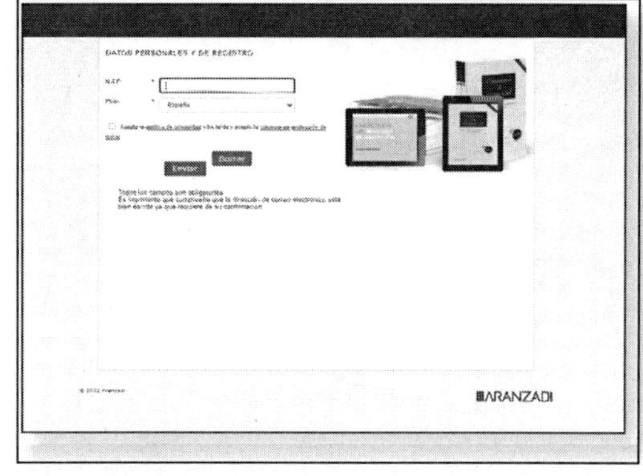

A continuación pulsa enviar.

Si te has registrado anteriormente en OnePass, en la siguiente pantalla se te pedirá que introduzcas el NIF asociado al correo electrónico.

Finalmente, te aparecerá un mensaje de confirmación y recibirás un correo electrónico confirmando la disponibilidad de la obra en tu biblioteca.

Si es la primera vez que te registras en **OnePass,** deberás cumplimentar los datos para crear tu cuenta y poder acceder a tu libro electrónico.

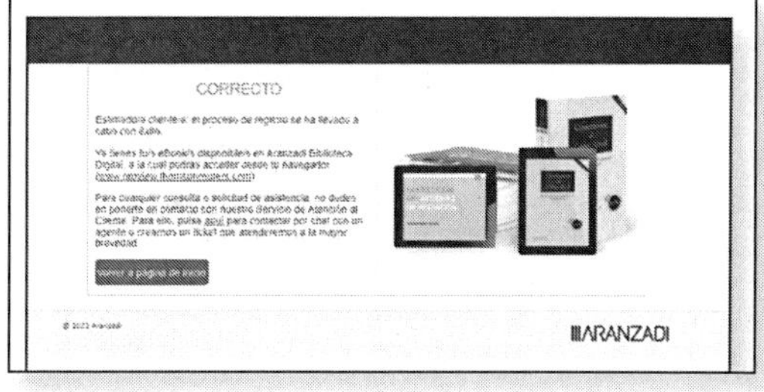

- Los campos **"Nombre de usuario"** y **"Contraseña"** son los datos que utilizarás para acceder a las obras que tienes disponibles a través del navegador en la ruta www.proview.thomsonreuters.com

Servicio de Atención al Cliente

Ante cualquier incidencia en el proceso de registro de la obra no dudes en ponerte en contacto con nuestro Servicio de Atención al Cliente. Para ello accede a nuestro Portal Corporativo y una vez allí en el apartado del Centro de Atención al Cliente selecciona la opción de Acceso a Soporte para no Suscriptores (compra de Publicaciones).